A C S S Y M P O S I U M S E R I E S **525**

Bioactive Volatile Compounds from Plants

Roy Teranishi, EDITOR
Agricultural Research Service
U.S. Department of Agriculture

Ron G. Buttery, EDITOR
Agricultural Research Service
U.S. Department of Agriculture

Hiroshi Sugisawa, EDITOR
Kagawa University

Developed from a symposium sponsored
by the Division of Agricultural and Food Chemistry
at the 203rd National Meeting
of the American Chemical Society,
San Francisco, California
April 5–10, 1992

American Chemical Society, Washington, DC 1993

Library of Congress Cataloging-in-Publication Data

Bioactive volatile compounds from plants /   Roy Teranishi,
Ron G. Buttery, Hiroshi Sugisawa

    p.   cm.—(ACS Symposium Series, 0097–6156; 525).

    "Developed from a symposium sponsored by the Division of
Agricultural and Food Chemistry at the 203rd National Meeting of
the American Chemical Society, San Francisco, California, April
5–19, 1992."

    Includes bibliographical references and indexes.

    ISBN 0–8412–2639–3

    1. Essences and essential oils—Congresses. 2. Plant bioactive
compounds—Congresses. 3. Plants—Analysis—Congresses.

    I. Teranishi, Roy, 1922–   . II. Buttery, Ron G. III. Sugisawa,
Hiroshi, 1928–   . IV. American Chemical Society. Division of
Agricultural and Food Chemistry. V. American Chemical Society.
Meeting (203rd: 1992: San Francisco, Calif.) VI. Series.

QK898.E75B56   1993
660'.6—dc20   93–18309
                                                       CIP

The paper used in this publication meets the minimum requirements of American National
Standard for Information Sciences—Permanence of Paper for Printed Library Materials, ANSI
Z39.48–1984.   ∞

Foreword

THE ACS SYMPOSIUM SERIES was first published in 1974 to provide a mechanism for publishing symposia quickly in book form. The purpose of this series is to publish comprehensive books developed from symposia, which are usually "snapshots in time" of the current research being done on a topic, plus some review material on the topic. For this reason, it is necessary that the papers be published as quickly as possible.

Before a symposium-based book is put under contract, the proposed table of contents is reviewed for appropriateness to the topic and for comprehensiveness of the collection. Some papers are excluded at this point, and others are added to round out the scope of the volume. In addition, a draft of each paper is peer-reviewed prior to final acceptance or rejection. This anonymous review process is supervised by the organizer(s) of the symposium, who become the editor(s) of the book. The authors then revise their papers according to the recommendations of both the reviewers and the editors, prepare camera-ready copy, and submit the final papers to the editors, who check that all necessary revisions have been made.

As a rule, only original research papers and original review papers are included in the volumes. Verbatim reproductions of previously published papers are not accepted.

M. Joan Comstock
Series Editor

Contents

Preface

VOLATILE COMPOUNDS FROM PLANTS, especially from flowers and fruit, have interested humans since ancient times. In perfumery and flavor chemistry studies, tens of thousands of compounds have been identified and correlated to characteristic odors.

Studies of volatile compounds in relation to humans have extended from simple olfactory research to examinations of medicinal properties and physiological and psychological effects. Plant–insect research is now under way to correlate volatile compounds from plants to insect responses, such as attraction or repulsion.

With the advent of the condenser, humans began isolating essential oils. Because of harsh distillation conditions, some essential oils do not have the odor qualities of fresh, living plants. Modern headspace isolation methods and modern analytical instruments, which require very small amounts of material, now permit researchers to sample trace volatile compounds emitted by living plants. The goal of perfumery and flavor chemists is to identify the volatile compounds that are responsible for the aromas of living flowers and fresh fruits and vegetables. Studies that compare results from various isolation methods are described in this book. Such data are very important in insect–plant research because insects respond to compounds emanating from living plants and usually not to those found in most distilled essential oils.

In this book, internationally known perfumery and flavor chemists from industry, government, and academia present the latest findings of volatile bioactive compounds from plants.

ROY TERANISHI
Agricultural Research Service
U.S. Department of Agriculture
800 Buchanan Street
Albany, CA, 94710

RON G. BUTTERY
Agricultural Research Service
U.S. Department of Agriculture
800 Buchanan Street
Albany, CA, 94710

HIROSHI SUGISAWA
Professor Emeritus
Department of Bioresource Science
Kagawa University, Miki-cho
Kagawa 761–07, Japan

August 19, 1992

We dedicate this volume to Peace

—RT, RGB, HS—

Chapter 1

Bioactive Volatile Compounds from Plants
An Overview

Roy Teranishi and Saima Kint

Western Regional Research Center, Agricultural Research Service,
U.S. Department of Agriculture, 800 Buchanan Street, Albany, CA 94710

Volatiles from plants have been used for perfumes and incense since ancient times. Today, the perfume industry incorporates extensive research on plant volatiles. Flavor chemists also study and use volatiles from plants to improve food quality. The vast information accumulated by perfume and flavor chemists forms an excellent base of knowledge for studying plant-insect interactions. The following is an overview on subsequent chapters which discuss some of the latest methods and equipment in sample preparation, analyses and identifications of volatiles from various flowers, leaves, and fruits. Also, activities other than olfactory responses, such as bacterial growth inhibition and increase or decrease of motor activity, are briefly reviewed.

Plants, either directly or indirectly, are probably most important to man as a source of food. Man also depends on plants for sources of medicine (1). Although the human being no longer utilizes chemical communication for survival, in ancient times, volatiles from some plants were used as incense and perfumes and were highly prized, valued as precious as gold. For example, the Wise Men who came to worship the newborn Jesus brought gold, frankincense, and myrrh as gifts (2). Incense was burned on altars to the gods (3). Perfumery was probably started centuries before the Christian era (4). In modern times, perfumery has become a large industry, and the intricate art, science, and technology are described in a recent book (5). The chemistry of perfumes is complex and elegant (5, 6). Early studies on terpenes established a better understanding of chemical bonding and molecular rearrangements in volatile compounds (7).

Chemical Communications

Most people are pleased with the fragrance of flowers, the smell of grass, the aroma of a pine forest. This human response is an indirect use of odors to convey a message and is the conceptual basis of the perfume industry. Chemical communication, however, does play an explicit role in lower animal forms: in search for food and mates, and in the avoidance of predators and danger. In the last few decades, much has been learned about chemical communications during insect-insect interactions and about pheromones which evoke responses among insects of the same species (8-16). Now research emphasis is directed towards insect-plant interactions (17, 18). In order to study and identify the volatiles from plants which evoke various responses in insects, we must turn to knowledge gained from perfume and flavor studies. Perfume chemists have identified and catalogued tens of thousands of compounds; flavor chemists, about 5,000 compounds.

Experimental

Some essential oils have aromas which seem to have no relationship to the plants from which they were obtained. In recent years, studies have attempted to identify the numerous compounds emanated from flowers and other parts of living plants (19). Some recent advances in capturing volatiles from living plants are covered in this book. To investigate plant-insect interactions, information from such studies is absolutely necessary. Correlating insect responses to specific compounds requires the knowledge of volatiles released by living plants. It is now known that plants release a range of concentrations of various compounds during different times of the day (20) and at different stages of fruit maturity (21). Upon fruit maceration, some compounds are released completely in a matter of minutes (22, 23).

Steam distillation and extraction methods have yielded essential oils from many plants and have generated much useful information. Advances in instrumentation and analytical methods now permit spectral analyses of very small quantities. Headspace volatiles were analyzed as early as 30 years ago by gas chromatography (24). Today, the combination of gas chromatography/mass spectrometry is used routinely. It is now possible to obtain infrared (25) and nuclear magnetic resonance (26) data on material separated with fused silica capillary columns. Results obtained with state-of-the-art separation techniques and instrumentation for sample preparation and analyses are discussed in this book. The differences in aromas obtained by different isolation methods are compared.

Discussion

Until recently the predominant sources of volatiles for perfumes have been essential oils and synthetic materials. It has been shown that there are considerable differences between volatiles from living and picked material from a certain plant and the essential oils obtained from the same plant by distillation or extraction (19). Because it is the dream of some perfumers to capture the freshness of living flowers, there is a growing interest in headspace analyses of living flowers and in vacuum headspace analyses of freshly cut flowers. Some advantages and disadvantages of such methods are discussed in this book.

Citrus fruit juices are freshly squeezed onto some foods for tartness and for fresh aroma. Extensive effort has been made to identify the compounds which contribute to this fresh quality. Again, the advent of new methods and equipment permit the investigation of such labile qualities. Studies of several citrus flowers and fruit are described in several chapters of this book.

Not only are the differences in volatiles from flowers from different species catalogued and described, but analytical methods have also been developed to such an extent that varietal differences can be characterized. A considerable data base is being developed for chemotaxonomy.

The use of attractive aromas from plants is not restricted to human beings. The age of indiscrimate use of non-selective, long lasting poisons for control of insects and weeds is slowly coming to an end after the plea by Rachel Carson (27) to use methods more compatible with the environment in which we live. "Through all these new, imaginative, and creative approaches to the problem of sharing our earth with other creatures there runs a constant theme, the awareness that we are dealing with life --- with living populations and all their pressures and counter-pressures, their surges and recessions. Only by taking account of such life forces and by cautiously seeking to guide them into channels favorable to ourselves can we hope to achieve a reasonable accommodation between the insect hordes and ourselves..." (27).

The use of insect attractants, not just pheromones, is merely beginning. We must find subtle ways to control insects by studying the plant-insect interactions and then intercede at vulnerable steps. To do so, we must study how specific plant volatiles modify the behavior of specific insects. The information accumulated by perfume and flavor chemists serves as a data base for such studies. Information gained from the biosynthesis of terpenes which attract insects is also very valuable.

In the world of medicine, it is well known that strong drugs may have detrimental side effects. Perhaps we should rely less on the pharmaceutical industry and use more subtle methods correlated with plant science. The inhibition of bacterial growth by some terpenoid compounds is well known.

Plant volatiles have been used since ancient times for soothing and promoting sleep. Now there is quantitative data on some volatile compounds which decrease or increase motor activity. Some interesting information in aromatherapy is given in this book.

It would take many volumes to cover all of the bioactivity evoked by plant volatiles. In this volume, we present some of the exciting topics under study on volatiles from flowers and other parts of plants by perfume and flavor chemists, together with the methods developed for isolating volatiles from living plants, and a very brief look at some bioactive aspects of such volatiles. It is hoped that this information will serve as a beginning for plant-insect interaction studies.

Literature Cited

1. Shultes, R. E. In *Plants in the Development of Modern Medicine*; Swain, T., Ed.; Harvard University Press: Cambridge, MA, **1972**; pp 103-124.
2. Matthew. In the New Testament, *The Holy Bible*, American Revision; Thomas Nelson & Sons: New York, NY, **1901**; 2, pp 11.

3.   Stoddart, D. M. *The Scented Ape*; Cambridge University Press: Cambridge, MA, **1990**; pp 168-206.

4.   Roudnitska, E. In *Perfumes: Art, Science and Technology*; Müller, P. M.; Lamparsky, D., Eds.; Elsevier Applied Science: London, **1991**; pp 3-48.

5.   *Perfumes: Art, Science and Technology*; Müller, P. M.; Lamparsky, D.; Eds.; Elsevier Applied Science: London, **1991**; pp 658.

6.   Ohloff, G. *Riechstoffe und Geruchssinn*; Springer-Verlag: New York, NY, **1990**; pp 233.

7.   Ruzicka, L. In *Perspectives in Organic Chemistry*; Todd, A.; Ed.; Interscience: New York, NY, **1956**; pp 265-314.

8.   Jacobson, M. *Insect Sex Pheromones*; Academic Press, New York, **1972**; pp 1-382.

9.   *Pheromones*; Birch, M. C., Ed.; North-Holland Publishing Company: Amsterdam, **1974**; pp 1-495.

10. *Introduction to Insect Pest Management*; Metcalf, R. L.; Luckmann, W. H.; Eds.; John Wiley and Sons: New York, NY, **1975**; pp 1-587.

11. *Chemicaontrol of Insect Behavior: Theory and Application*; Shorey, H. H.; McKelvey, J. J., Jr.; Eds.; John Wiley and Sons: New York, NY, **1977**; pp 1-414.

12. *Insect Suppression with Controlled Release Pheromone Systems*; Kydonieus, A. F.; Beroza, M.; Zweig, G.; Eds.; CRC Press, Inc.: Boca Raton, FL, **1982**; *vol 1*, pp 1-274, *vol 2*, pp 1-312.

13. *Techniques in Pheromone Research*; Hummel, H. E.; Miller, T. A.; Eds.; Springer-Verlag, New York, NY, **1984**; pp 1-446.

14. *CRC Handbook of Natural Pesticides, Volume IV, Pheromones, Part A*; Morgan, E. D.; Mandava, N. B.; Eds; CRC Press, Inc.: Boca Raton, FL, **1988**; pp 1-203.

15. *CRC Handbook of Natural Pesticides, Volume IV, Pheromones, Part B*; Morgan, E. D.; Mandava, N. B.; Eds.; CRC Press, Inc.: Boca Raton, FL, **1988**; pp 1-291.

16. *Handbook of Insect Pheromones and Sex Attractants*; Mayer, M. S.; McLaughlin, J. R.; Eds.; CRC Press, Inc. Boca Raton, FL, **1991**; pp 1-1083.

17. *Semiochemicals: Their Role in Pest Control*; Nordlund, D. A.; Jones, R. L.; Lewis, W. J.; Eds.; John Wiley and Sons: New York, NY, **1981**; pp 1-306.

18. *CRC Handbook of Natural Pesticides: Volume VI, Insect Attractants and Repellents*; Morgan, E. D.; Mandava, N. B.; Eds.; CRC Press, Inc.: Boca Raton, FL, **1990**; pp 1-249.

19. Mookherjee, B. D.; Wilson, R. A.; Trenkle, R. W.; Zampino, M. J., Sands, K. P.; *Flavor Chemistry: Trends and Developments*; ACS Symposium Series 388; American Chemical Society: Washington, DC, **1989**; pp 176-187.

20. Loper, G. M.; Berdel, R. L. *Crop Science*; **1978**, *vol 17*, pp 447-452.

21. Engel, K.-H.; Ramming, D. W.; Flath, R. A.; Teranishi, R. *J. Agric. Food Chem.*, **1988**, *vol 36*, pp 1003-1006.

22.   Drawert, F.; Kler, A.; Berger, R.G. *Lebensm.-Wiss. u.-Technol.* **1986**, *vol 19*, pp 426-431.

23.   Buttery, R. G.; Teranishi, R.; Ling, L. C. *J. Agric. Food Chem.*; **1987**, *vol 35*, pp 540-544.

24.   Teranishi, R.; Buttery, R. G. *Volatile Fruit Flavors;* International Federation of Fruit Juice Producers Symposium, Bern, Switzerland; Juris-Verlag: Zürich, **1962**; pp 257-266.

25.   Williams, A.A.; Tuchnott, O. G.; Lewis, M. J.; May, H.; Wachter, L. In *Flavor Science and Technologie;* Martens, M.; Dalen, G. A.; Russwurm, Jr., H.; Eds.; John Wiley & Sons: Chichester, **1987**, pp 259-270.

26.   Etzweiler, F. *J. High Res. Chrom. & Chrom. Comm.* **1988**, pp 449-456.

27.   Carson, R. *Silent Spring;* Houghton Miffin Co: Boston, MA, **1962**; 368 pp.

RECEIVED August 19, 1992

BIOGENESIS AND BIOCHEMISTRY

Chapter 2

Conifer Monoterpenes

Biochemistry and Bark Beetle Chemical Ecology

Mark Gijzen, Efraim Lewinsohn, Thomas J. Savage,
and Rodney B. Croteau

Institute of Biological Chemistry, Washington State University,
Pullman, WA 99164–6340

Monoterpenes comprise about half of the oleoresin produced by pines, spruces, firs and related conifer species, with the remainder largely diterpenoids. The volatile monoterpenes are phytoprotective agents, defending the tree from herbivore and pathogen attack; however, many species of tree-killing bark beetles are attracted to monoterpene compounds emitted from conifers and can utilize components of the host oleoresin to manufacture pheromones for promoting aggregation. These same volatile products may also serve as attractants for beetle predators and parasitoids. This complex relationship between conifer host, beetle pest, and beetle predator is briefly reviewed, and the biochemistry of oleoresin monoterpenes is described with the view to manipulating this process to improve tree resistance.

Conifer Oleoresin

The conifers are an ancient group of woody plants that date from at least 200 million years ago. There are some 600 extant species of conifers classified in six different families (1). Of these, the Pinaceae are the most abundant and widespread, especially in the northern hemisphere, and they include many species that are prolific producers of oleoresin (2), a complex mixture of roughly equal amounts of monoterpenes (turpentine) and diterpenoid resin acids (rosin) (3-5) (Fig. 1). Conifer oleoresin, also called simply resin or pitch, may be synthesized and stored in an intricate system of resin passages or ducts as in the spruces (*Picea*) and pines (*Pinus*) (6), or in disperse resin cells or multicellular blisters as in the true firs (*Abies*) (7,8). In either case, the release of this sticky and viscous substance upon tissue injury presents a formidable physical and chemical barrier that serves to protect the tree against pests and pathogens, and ultimately seals the wound.

Oleoresin is toxic and/or repellent to many pests and pathogens, and the survival of a tree under attack may depend upon the production and mobilization of copious amounts of this material (9). The toxicity of oleoresin monoterpenes towards bark beetles, one of the most serious insect pests of conifer species, and the inhibitory effects of these compounds upon the growth of pathogenic fungal symbionts of the beetles have been well documented. Death of the southern pine

0097–6156/93/0525–0008$06.00/0

beetle (*Dendroctonus frontalis*) and the engraver *Ips calligraphus* resulting from exposure to oleoresin monoterpene vapors has been demonstrated (10). The monoterpenes of grand fir (*Abies grandis*) in vapor form cause significant mortality to fir engraver beetles (*Scolytis ventralis*) within 4 h of exposure and they inhibit the growth of the symbiotic fungus (*Trichosporium symbioticum*) (11). Tested separately, all of the individual monoterpenes of grand fir oleoresin are toxic to the beetle and fungus, with only minor differences in toxicity among the different olefin isomers (11). Many other conifer pathogens, including virulent fungi of the genus *Ceratocystis*, are sensitive to oleoresin monoterpenes, either administered in vapor form or included in the growth medium (12-16).

In addition to the constitutive oleoresin accumulated in the specialized secretory systems, conifers may also synthesize new oleoresin in response to wounding and infection, and, with time, develop auxiliary resin producing structures called traumatic resin ducts (7,8,17). Wound- or pathogen-induced oleoresin, also called secondary oleoresin, may differ in composition from constitutive oleoresin (also called preformed, stored or primary oleoresin), especially with regard to the monoterpene components. For example, induced oleoresin accumulated at wound sites on white fir (*A. concolor*) stem contains a much higher proportion of α–pinene relative to preformed resin stored in blisters (18). Similarly, α–pinene was more prevalent in oleoresin extracted from fungus infected (*T. symbioticum*) grand fir stem tissue than from blister oleoresin (19). However, in another study with grand fir it was found that the α–pinene content of the oleoresin decreased in response to infection with *T. symbioticum* (20). The difficulty of distinguishing constitutive oleoresin from induced oleoresin by chemical analysis, coupled to the considerable volatility of the monoterpenes on exposure to the atmosphere, may explain this apparent contradiction. It should be noted in this connection that the tissues surrounding the site of injury or infection generally become saturated with oleoresin and phenolic materials, forming a "reaction zone" (Fig. 2). Evaporation of the monoterpenes that serve as a solvent, and subsequent crystallization and/or oxidative polymerization of the diterpene resin acids (Fig. 1) so deposited, leads to wound sealing by the formation of a tough, plastic protective coating (3).

Bark Beetles

The bark beetles (Coleoptera:Scolytidae) comprise a group of insects that live on the wood, bark, or roots of trees. Although most of these species feed on dead tissue, a few genera can attack and kill living trees (20,21). These aggressive types, primarily species of *Dendroctonus, Ips*, and *Scolytus*, are among the most destructive insect pests of conifer forests in North America (9). Outbreaks may be scattered and isolated, or spread over hundreds of square kilometers. These bark beetles display a high degree of host specialization; a given species will usually attack only one, or perhaps two, species of conifers (22). This selectivity likely reflects the extended period of co-evolution of beetles and conifers. Indeed, scolytid-like gallery patterns have been observed in 200 million-year-old petrified wood samples, and beetle specimens entrapped in fossilized oleoresin (amber) of pine trees that grew nearly 40 million years ago have also been described (22).

Although details of the bark beetle life cycle vary with the species and geographic location, there are general features that are common to nearly all types (21). A brief dispersal phase is characterized by emergence of adult beetles from the dead host and flight to a new, living host. The dispersing beetles aggregate and colonize the new host *en masse* (Fig. 3A), since in most species the tree must be killed to insure successful reproduction. Once a suitable host has been selected, beetles bore through the bark and establish brood galleries in the cambium (Fig. 3C), at which time the host tissue is inoculated with a pathogenic fungus carried by

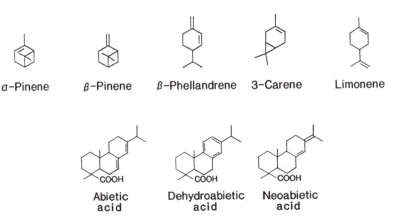

α-Pinene   β-Pinene   β-Phellandrene   3-Carene   Limonene

Abietic
acid

Dehydroabietic
acid

Neoabietic
acid

Figure 1. Representative monoterpenes and diterpene resin acids of conifer oleoresin.

Figure 2. Reaction zone surrounding a fir engraver (*S. ventralis*) bore hole on grand fir (*A. grandis*). Photograph taken by the authors near La Grande, Oregon.

Figure 3. (A) Mass attack by mountain pine beetles (*D. ponderosae*) on a lodgepole pine bole (*P. contorta*). Each white spot on the trunk represents a beetle entry point. (B) Close up of an unsuccessful attempt by *D. ponderosae* to colonize ponderosa pine (*P. ponderosa*) in which the beetle has been "pitched out". (C) Beetle galleries (arrow) below the bark of a colonized ponderosa pine. Photographs taken by the authors near Bend, Oregon.

the beetles in specialized structures (mycangia). Tree killing is thought to result from girdling of the tree by beetle excavation, disruption of xylem flow by fungal invasion, and cell death via fungal toxin production. Larvae develop from eggs laid along the gallery walls and they mine tunnels of their own as they feed on the stem tissue. The larvae then pupate and the adult beetles bore exit holes to disperse and find new hosts. All stages of the life cycle, host identification, aggregation, infestation, and reproduction, can be significantly influenced by the quantity and composition of the oleoresin produced by the conifer host.

Host Location and Selection

The initial phase of bark beetle attack, location of a suitable host, is driven by olfactory, gustatory, and visual signals perceived by the searching beetle, although beetle dispersion, itself, is a seemingly random process that is influenced by environmental conditions (23-25). Newly felled trees or trees damaged by lightning or wind may be especially attractive targets (26), as are baited traps containing cut log sections or turpentine extracts (27,28). The presence of freshly exposed oleoresin as a result of host injury usually results in the convergence of any free-flying beetles in the vicinity. Even intact trees emit small quantities of oleoresin monoterpenes, which may be a sufficient attractant to permit beetles to distinguish host from non-host species. (29). Trees that are stressed by drought, pollution or other physiological disturbance are more susceptible to bark beetle invasion; however, there is no evidence that such trees are more attractive to host-seeking beetles (30).

The paradox, that beetles are attracted to material that is toxic, illustrates the complex evolutionary interaction between conifer hosts and herbivores. For the bark beetles, the benefit of utilizing specific volatile cues (oleoresin monoterpenes) to locate a host apparently outweighs the disadvantage of attraction to these toxic substances. Selection over many generations has produced beetles that are adapted to a particular host or phenotype, and conifers that can muster multiple, diverse chemical defenses (31).

Upon landing on the tree trunk, the beetle may commence boring into the new host based on textural and compositional features of the bark (32) or, alternatively, deem the tree unsuitable and resume flight. Following penetration, beetles may still abandon the bore or, alternatively, seek to establish residence in the cambium. It is clear that additional cues, beyond those employed in host location and identification, are utilized in host selection (32). If, in their efforts to establish a gallery in the selected host, the beetles sever resin passage or duct systems, they may be overcome by exuded oleoresin and die (23,33). This "pitching-out" of beetles by rapid oleoresin flow to the site of injury (Fig. 3A,B) is considered to be among the most important of chemical defense mechanisms in repelling the initial beetle invasion (34-36). The efficacy of this process is primarily dependent upon the composition (37,38) and amount (39,40) of oleoresin, and the ability to mount such a successful counteroffensive may be compromised if the tree is physiologically weakened (26,30,40,41). Only conifer species with substantial constitutive oleoresin stores, such as pines and spruces, can rely on this pitching-out strategy.

Beetles that attack conifer species that, like the true firs, lack resin duct or passage systems, may nonetheless be overcome by the exudation of blister oleoresin (42). Wound-inducible oleoresinosis presents a second line of defense with which the invading beetles must contend. This enhanced biosynthesis of oleoresin in the immediate vicinity of a wound or infection site (43-45) seems especially important in conifer species that do not accumulate large reserves of constitutive oleoresin within highly developed secretory systems (9). It has recently been shown that the induced production of oleoresin in grand fir can be impaired if the tree is water stressed (46), thereby providing a rationale for the observation that weakened trees are more likely to succumb to beetle attack.

Aggregation and Colonization

The observation that beetles tend to aggregate on individual trees was noted early in studies of bark beetle ecology (47). The main advantage of this focused attack is that the tree is unable to deal with so many intruders simultaneously, and hence is overpowered by sheer force of numbers (Fig. 3A) (33,48). This concentration of attackers has been variously attributed to chemical signals arising from the wounded tree, from the initial beetle invaders, or from the infecting fungus (49-51). The most important of these signals are the beetle-produced pheromones, many of which are synthesized by the beetle from substrates present in the host oleoresin.

Pheromones are substances that, secreted by one organism, influence the behavior or physiology of other organisms of the same species. In the Scolytidae, these semiochemicals may play many different roles (52,53). Aggregation pheromones attract individuals of both sexes to a host tree, whereas spacing or antiaggregation pheromones insure that colonization does not achieve a level where competition between beetles becomes too intense (50). Certain pheromones can also induce reproductive activity or stridulation. Specific aggregation pheromones may be active as such, or synergized by combination with other pheromones or host resin monoterpenes. Additionally, the emitted pheromones may be perceived by other species of bark beetles or by beetle predators and parasitoids. Allomones are signals that, detected by another species, serve to benefit the emitter, whereas kairomones have the opposite effect and serve the receiver. Thus, an allomone may define territory and repel a competing insect species, while a kairomone typically aids predator and parasite species in locating prey.

The first scolytid aggregation pheromones identified were ipsdienol, ipsenol, and *cis*-verbenol (Fig. 4), isolated from the frass of *Ips paraconfusus* (54). The oxygenated myrcene derivatives, ipsdienol and ispsenol, seem to be important in the communication systems of all *Ips* species examined thus far (50). As with many beetle pheromones, ipsdienol is a chiral compound. The (+)-enantiomer is an aggregation pheromone of *I. paraconfusus*, whereas (-)-ipsdienol acts as an aggregation inhibitor. The opposite is true for the competing species *Ips pini* (the pine engraver) for which (-)-ipsdienol is aggregation-attractive and (+)-ipsdienol is aggregation-inhibitory (55). The selective use of different ipsdienol enantiomers by *I. paraconfusus* and *I. pini* is thought to be important in regulating interspecific competition between these two co-occurring species. The chiral preferences of *I. pini*, however, may depend upon geographical location, since eastern populations of *I. pini* are attracted most strongly by a mixture of both enantiomers (56,57). Raffa and Klepzig (58) have suggested that the chiral specificity of beetle pheromones may evolve as a response to predator recognition. These workers showed that two species of predatory beetles that feed upon *I. pini* are attracted by the mixture of ipsdienol antipodes produced by their prey; however, the stereoisomer mixture that was optimally attractive for the predator species was different from that which was most attractive to the prey (58). Allomonal and kairomonal effects may therefore impose strong selective pressures favoring the use of one or the other enantiomer of a pheromone compound.

The oxygenated α–pinene derivatives, *cis*- and *trans*-verbenol, and verbenone (Fig. 4), are also important monoterpene pheromones. The enantiomeric composition of the α–pinene of conifer host oleoresin is significant in this regard because it defines the stereoisomer(s) of verbenol that can be produced by the beetle. Thus, in *I. paraconfusus*, (-)-α–pinene serves as the precursor of (+)-*cis*-verbenol and (-)-verbenone, whereas (+)-α–pinene leads to (+)-*trans*-verbenol and (+)-verbenone (Fig. 4) (59). As with ipsdienol and ipsenol, the various verbenol

| Myrcene | Ipsdienol | Ipsenol |

| (−)-α-Pinene | (+)-cis-Verbenol | (−)-Verbenone |

 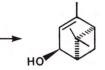

| (+)-α-Pinene | (+)-trans-Verbenol | (+)-Verbenone |

Figure 4.   Structures of selected monoterpenoid pheromones and their corresponding olefin precursors.

stereoisomers differ greatly in their biological activity. In the field, *I. typographus* is more strongly attracted to (+)-*cis*-verbenol than to other verbenol stereoisomers (57). Francke and Vite (60) studied the time-dependent release of α–pinene-derived pheromones produced by *I. typographus* and found that, during the initial stages of colonization, there is a much greater proportion of the attractive components, *cis*- and *trans*-verbenol, in the emitted mixture compared to the inhibitory compound verbenone. These investigators suggest that the time-dependent shift in pheromone production from verbenol to verbenone serves to regulate attack density. In many *Dendroctonus* species, *trans*-verbenol is a strong attractant that often acts synergistically with other pheromones or host monoterpenes (61,62). Indeed, *trans*-verbenol is ineffective as an attractant for *D. ponderosae* (the mountain pine beetle) unless presented in combination with host volatiles (25,63). The identification of bioactive components of the host volatile complex can be complicated by the spontaneous oxidation of α–pinene (and perhaps other monoterpenes) under ambient conditions (64).

Less common monoterpene pheromones include *cis*- and *trans*-4-thujanol, terpinen-4-ol and α–terpineol (60). These compounds could arise in the process of host oleoresin biosynthesis, or they may be derived by subsequent oxygenation by the beetle of a thujane or *p*-menthane monoterpene, respectively. Several non-monoterpenoid pheromones, including the bicyclic ketals, brevicomin, frontalin, and multistriatin, have been identified, and their biological activities described (50,52,53,60,65). These compounds can interact synergistically with monoterpenoid semiochemicals.

Pheromone biosynthesis by bark beetles has been the subject of some debate. It has been shown that free-living fungal symbionts and gut bacteria are able to oxidize monoterpenes such as α–pinene and myrcene to the corresponding pheromones (49,51,66-68). By contrast, there is a report on the production of ipsdienol and ipsenol in *I. paraconfusus*, apparently in the absence of the myrcene precursor, from which the authors suggest that bark beetles may have the capacity to synthesize these compounds *de novo* (69). However, the influence of oleoresin composition on pheromone production by beetles (59), and the *in vivo* conversion of labelled monoterpenoid precursors to the corresponding pheromones (70), support the more widely held belief that terpenoid pheromones are derived from host plant monoterpenes. Regardless of origin, the developmental and sex-specific release of pheromones indicates that beetles can control the production of monoterpenoid semiochemicals.

The dependence of bark beetle behavior on signal compounds derived from monoterpenes of host oleoresin presents an interesting evolutionary adaptation. It has been proposed that oxidative modification of host monoterpenes by beetles may have arisen as a detoxification mechanism (60). The subsequent recruitment of these monoterpenoid derivatives to serve as semiochemicals seems a natural development. A further extension of the signalling complex beyond the original defensive compounds of the oleoresin and their expropriation as bark beetle pheromones is represented by the use of these same compounds by bark beetle predators and parasitoids to locate their respective prey and hosts (71). Whereas some parasitic wasps are attracted to their beetle hosts by the emitted pheromones, many parasitoids of this type are more strongly attracted to tree volatiles and, thus, may attack bark beetles on one type of conifer but not attack the same species of bark beetles on another type of tree (71). Insect predators of bark beetles also respond to host pheromones (and to conifer volatiles), but the specificity of these responses is not very well understood in part because these predators tend to feed on a wide range of prey and may exploit a broad set of chemical signals. A particularly illustrative example of this complex interaction between three trophic levels is the association between *Picea excelsa* (Norway spruce), the European spruce bark beetle (*Dendroctonus micans*) and the predatory beetle *Rhizophagus grandis.* Oleoresin monoterpenes (α–pinene, β–pinene, and others) attract both bark beetles and predator beetles (72), whereas the beetle pheromones (*cis*- and *trans*-verbenol and verbenone) seemingly serve as oviposition stimulants for *R. grandis* (73). Interestingly, *R. grandis* adults are almost totally resistant to contact toxicity with spruce monoterpenes at doses equivalent to half their own weight (74).

Oleoresin Biosynthesis

Terpenoids (isoprenoids) constitute the largest family of natural products, and they are derived from the ubiquitous acetate-mevalonate pathway (75). Thus, three molecules of acetyl CoA are condensed to 3S-hydroxymethylglutaryl CoA that is reduced to 3R-mevalonic acid. Subsequent, stepwise phosphorylation and decarboxylative elimination yields isopentenyl pyrophosphate, and isomerization of the latter to dimethylallyl pyrophosphate provides the two basic C_5 precursors to all other intermediates of the pathway (Fig. 5). Condensation of isopentenyl pyrophosphate and dimethylallyl pyrophosphate catalyzed by prenyltransferase leads to geranyl pyrophosphate, the immediate precursor of the monoterpenes, while further elongation provides farnesyl pyrophosphate (C_{15}) and geranylgeranyl pyrophosphate (C_{20}), the immediate precursors of the sesquiterpenes and diterpenes, respectively. Although there are many examples of isoprenoid derivatives that serve essential structural or developmental roles in plants (e.g., phytosterols and gibberellins), most terpenoids are synthesized as agents of communication or defense (76). The production of large quantities of oleoresin in conifers is an ancient and seemingly successful example of the latter.

   The odor-bearing monoterpenes of the volatile fraction of conifer oleoresin (turpentine) are principally olefins; however, the occurrence of oxygenated monoterpenes, at least in small quantities, is not uncommon. The monoterpenes are synthesized directly from geranyl pyrophosphate by enzymes termed monoterpene synthases (or, more commonly, monoterpene "cyclases" in the case of cyclic products, which are exceedingly common). The mechanism of the reaction catalyzed by the monoterpene cyclases has been examined in some detail (77), and it has been demonstrated that many monoterpene cyclases from conifers synthesize multiple products at the same active site (for example, (-)-α– and (-)-β–pinene, and (-)-α– and (-)-β–phellandrene) (78). Because of this unusual feature, the inheritance patterns of oleoresin monoterpene composition may be quite complex (79-81).

   The monoterpene biosynthetic capacity of different conifer species (as determined by cyclization activity per gram tissue or per mg protein in stem cell-free extracts) differs greatly, and generally correlates with both the quantity of oleoresin present and the anatomical sophistication of the oleoresin bearing structures of the stem (43). Comparison of constitutive monoterpene biosynthetic capacity with that of wound-induced monoterpene biosynthetic capacity, using saplings as a model system, revealed an interesting trend that implicates two fundamentally different defense strategies. Conifer species with well developed resin ducts, like the pines, were found to produce large amounts of oleoresin constitutively, and to respond to wounding by translocation of this material to the wound site (82). Other species such as grand fir, that possess limited constitutive biosynthetic activity and storage capacity, were shown to undergo dramatic localized enhancement of monoterpene biosynthesis within a few days after wounding (44). Mature grand fir trees in the forest setting respond in a similar manner by increasing monoterpene biosynthetic capacity at the wound site (46). As indicated previously, it is often difficult to distinguish, by analytical means, primary and secondary oleoresin exuded at a site of injury. The ability to measure biosynthetic capacity at the cell-free enzyme level

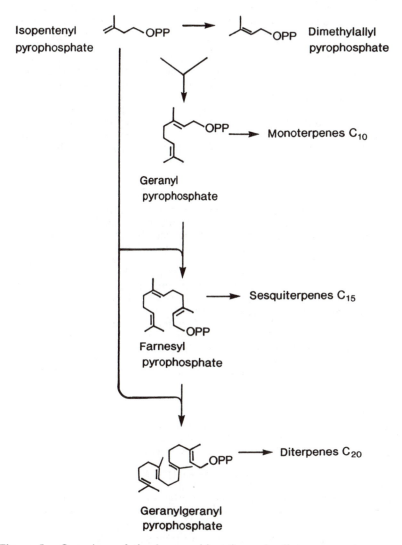

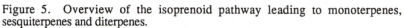

Figure 5. Overview of the isoprenoid pathway leading to monoterpenes, sesquiterpenes and diterpenes.

allows straightforward differentiation of constitutive and wound-induced processes (82,83).

The increase in monoterpene biosynthesis following wounding of grand fir is the result of the apparent enhancement of constitutive activities (including limonene cyclase) and the appearance of distinct, wound-inducible activities (pinene cyclase) (83,84). The major wound-inducible monoterpene cyclase of grand fir stem produces both (-)-α–pinene and (-)-β–pinene from geranyl pyrophosphate via a common cationic intermediate (78). This 62 kDa monomeric enzyme has been characterized, purified to apparent homogeneity and used to raise polyclonal antibodies in rabbits (85). Immunochemical methods were then employed to demonstrate that the pinene cyclase is synthesized *de novo* in response to wounding, and to show that this pinene cyclase is more closely related to other cyclases present in grand fir (that synthesize different monoterpene skeletal types) than it is to the pinene cyclases from related conifer species (86).

Examination of monoterpene biosynthesis in grand fir has demonstrated that wound-induced oleoresin production differs significantly from that of constitutive origin, notably in the increase in pinenes. Differential control of oleoresin formation, as in grand fir, is undoubtedly exploited by many other conifer types in the ongoing struggle against pests and pathogens.

Prospect

In contrast to the considerable experimental attention devoted to the biology and ecology of the tree-beetle-fungus-predator/parasitoid associations, the biochemistry of oleoresin formation in conifers has, until recently, received little attention. In spite of the experimental difficulties in working with conifers, this seeming neglect is nevertheless surprising since oleoresin seems to be the principal form of chemical defense and plays a major role in host selection, pheromone production, and attraction of bark beetle enemies. Individual monoterpenes may differ in repellency, toxicity and inhibitory effects on bark beetles and associated fungi, and in physical properties, such as solvent capacity and vapor pressure, that influence resin viscosity and crystallization rate. Therefore, considerable opportunity exists for altering tree resistance by modification of oleoresin content. Although the genetic manipulation of oleoresin composition would seem to offer an ideal means of biological disease control, this approach has yet to be exploited because oleoresinosis is very complex and regulation of this process, that underlies susceptibility and resistance, is still incompletely understood. Continued study of the enzymology and regulatory mechanisms of oleoresin biosynthesis, and the subsequent isolation and analysis of cDNAs encoding key steps of the pathway, should provide means for the enhanced protection of commercial timber species. The isolation of relevant cDNAs also should open another avenue of research using RFLP methods to analyze conifer oleoresin genes in relation to host-pathogen interactions.

The Plant Genome Project of the USDA National Research Initiative Competitive Grants Program has recognized the significance of conifers by including loblolly pine (*Pinus taeda*) as one of the four target plant species for genome sequencing. The USDA also supports genome mapping efforts for other conifer species.

Possible approaches for exploiting recombinant DNA methods in conifer defense include improving the speed and level of the oleoresin response at the critical early stages, or increasing the concentrations of compounds that are particularly toxic to invaders. More sophisticated strategies could involve altering oleoresin chemistry to modify attraction/repulsion and thereby confound host selection, to promote three-trophic level interactions that foster beetle predation or parasitism, or to abolish production of a pheromone precursor and thereby disrupt aggregation or mate selection.   In order to use any of these strategies, transformation and regeneration of conifer species is required. Although progress in this area of study has been slow, significant advances have been made. Recently, Huang and co-workers (87) reported the first transformation and regeneration of the conifer species *Larix decidua*. The susceptibility to *Agrobacterium tumefaciens* transfection has been demonstrated in many conifer genera including *Pinus*, *Picea* and *Abies* (88). Therefore, the groundwork for the production of transgenic conifers has been established. The apparent similarity of the protein catalysts responsible for terpene production [as well as the very large pool of these cyclase enzymes potentially available (77,82,83)], and their deployment in fundamentally different defense strategies, are design features of great biological interest and potential in this regard. Moreover, the exploitation of terpene cyclases offers a conceptually attractive and biorational approach for improving conifer defense by altering not only the mix of constitutive and induced oleoresinosis, but also the yield and composition of the oleoresin itself.

Acknowledgments

Research by the authors was supported in part by U.S. Department of Agriculture Grant 91-37302-6311, a grant from the McKnight Foundation, and by Project 0268 from the Washington State University Agricultural Research Center, Pullman, WA 99164.

Literature Cited

1. Scagel, R. F.; Bandoni, R. J.; Rouse; G. E.; Schofield, W. B.; Stein, J. R.; Taylor, T. M. C. *An Evolutionary Survey of the Plant Kingdom*; Wadsworth Publishing Company, Inc.: Belmont, CA, 1965; pp 491-524.
2. Penhallow, D. P. *A Manual of the North American Gymnosperms*; Athenaeum Press: Boston, MA, 1907; 374 pp.
3. Croteau, R.; Johnson, M. A. In *Biosynthesis and Biodegradation of Wood Components*; Higuchi, T., Ed.; Academic Press: New York, NY, 1985; pp 379-439.
4. Johnson, M. A.; Croteau, R. In *Ecology and Metabolism of Plant Lipids*; Fuller, G.; Nes, W. D., Eds.; ACS Symposium Series 325; American Chemical Society: Washington, DC; 1987; pp 76-92.
5. Norin, T. *Phytochemistry* **1972**, *11*, 1231-1242.
6. Werker, E.; Fahn, A. *Bot. J. Lin. Soc.* **1969**, *62*, 379-411.
7. Jeffrey, E. C. *Mem. Boston Soc. Nat. Hist.* **1905**, *6*, 1-37.
8. Bannan, M. W. *New Phytol.* **1936**, *35*, 11-47.
9. Berryman, A. A. *Bioscience* **1972**, *22*, 598-602.
10. Cook, S. P.; Hain, F. P. *J. Entomol. Sci.* **1988**, *23*, 287-292.
11. Raffa, K. F.; Berryman, A. A.; Simasko, J.; Teal, W.; Wong, B. L. *Environ. Entomol.* **1985**, *14*, 552-556.
12. Gibbs, J. N. *Ann. Bot.* **1988**, *32*, 649-665.
13. Krupa, S.; Nylund, J. E. *Eur. J. For. Path.* **1972**, *2*, 88-94.
14. Shrimpton, M. D.; Whitney, H. S. *Can. J. Bot.*, **1968**, *46*, 757-761.
15. Cobb, F. W. Jr.; Krstic, M.; Zavarin, E.; Barber, H. W. Jr. *Phytopathology* **1968**, *58*, 1327-1335.
16. De Groot, R.C. *Mycologia* **1972**, *64*, 863-870.
17. Fahn, A.; Zamski, E. *Israel J. Bot.* **1970**, *19*, 429-446.
18. Ferrell, G. T. In *Mechanisms of Woody Plant Defenses Against Insects. Search for Pattern*; Mattson, W. J.; Levieux, J.; Bernard-Dagan, C., Eds.; Springer-Verlag: New York, NY, 1988; pp 305-312.
19. Russell, C. E.; Berryman, A. A. *Can. J. Bot.* **1976**, *54*, 14-18.

20. Raffa, K. F.; Berryman, A. A. *Can. Entomol.* **1982**, *114*, 797-810.
21. Stark, R. W. In *Bark Beetles in North American Conifers: A System for the Study of Evolutionary Biology*; Mitton, J. B.; Sturgeon, K. B., Eds.; University of Texas Press: Austin, TX, 1982; pp 21-45.
22. Sturgeon, K. B.; Mitton, J. B. In *Bark Beetles in North American Conifers: A System for the Study of Evolutionary Biology*; Mitton, J. B.; Sturgeon, K. B., Eds.; University of Texas Press: Austin, TX, 1982; pp 350-384.
23. Ashraf, M.; Berryman, A. A. *Melandria* **1969**, *2*, 1-23.
24. Payne, T. L. *Z. Ang. Entomol.* **1983**, *96*, 105-109.
25. Billings, R. F.; Gara, R. I.; Hrutfiord, B. F. *Environ. Entomol.* **1976**, *5*, 171-179.
26. Furniss, M. M.; McGregor, M. D.; Foiles, M. W.; Partridge, A. D. *USDA For. Serv. Gen. Tech. Rep.*; INT-59, Intermt. For. and Range Exp. Stn.: Ogden, UT, 1979; 19 pp.
27. Wood, D. L.; Vité, J. P. *Contrib. Boyce Thompson Inst.* **1961**, *21*, 79-95.
28. Coulson, R. N.; Hennier, P. B.; Flamm, R. O.; Rykiel, E. J.; Hu, L. C.; Payne, T. L. *Z. Ang. Entomol.* **1983**, *96*, 182-193.
29. Rhoades, D. F. *Phytochemistry* **1990**, *29*, 1463-1465.
30. Wood, D. L.; Akers, R. P.; Owen, D. R.; Parmeter, J. R. In *Insects and the Plant Surface*; Juniper, B.; Southwood, R., Eds.; Edward Arnold Press: London, 1987; pp 91-103.
31. Edmunds, G. F.; Alstad, D.N. *Science* **1978**, *199*, 941-945.
32. Raffa, K. F. In *Mechanisms of Woody Plant Defenses Against Insects. Search for Pattern*; Mattson, W. J.; Levieux, J.; Bernard-Dagan, C., Eds.; Springer-Verlag: New York, NY, 1988; pp 369-390.
33. Christiansen, E.; Waring, R. H.; Berryman, A. A. *For. Ecol. Manage.*, **1987**, *22*, 89-106.
34. Shrimpton, M. D. In *Theory and Practice of Mountain Pine Beetle Management in Lodgepole Pine Forests*; Berryman, A. A.; Amman, G. D.; Stark, R. B.; Kibbee, D. L., Eds.; College of Forest Resources, University of Idaho: Moscow, ID, 1978; pp 64-76.
35. Cates, R. G.; Alexander, H. In *Bark Beetles in North American Conifers: A System for the Study of Evolutionary Biology*; Mitton, J. B.; Sturgeon, K. B., Eds.; University of Texas Press: Austin, TX, 1982; pp 212-263.
36. Cook, S. P.; Hain, F. P. In *Mechanisms of Woody Plant Defenses Against Insects. Search for Pattern*; Mattson, W. J.; Levieux, J.; Bernard-Dagan, C., Eds.; Springer-Verlag: New York, NY, 1988; pp 295-304.
37. Smith, R. H. In *Breeding Pest-Resistant Trees*; Gerhold, H. D.; McDermott, N. E.; Schreiner, E. J.; Winieski, J. A., Eds.; Pergamon Press: Oxford, 1966; pp 189-206.
38. Gollob, L. *Naturwissenschaften* **1980**, *67*, 409-410.
39. Hodges, J. D.; Elam, W. W.; Watson, W. F.; Nebeker, T. E. *Can. Entomol.* **1979**, *111*, 889-896.
40. Raffa, K. F.; Berryman, A. A. *Environ. Entomol.* **1982**, *11*, 486-492.
41. Waring, R. H.; Pitman, G. B. *Z. Ang. Entomol.* **1983**, *96*, 265-270.
42. Ferrell, G. T. *Can. Entmol.* **1983**, *115*, 1421-1428.
43. Lewinsohn, E.; Gijzen, M.; Savage, T. J.; Croteau, R. *Plant Physiol.* **1991**, *96*, 38-43.
44. Lewinsohn, E.; Gijzen, M.; Croteau, R. *Plant Physiol.* **1991**, *96*, 44-49.
45. Berryman, A. A. *Can. Entomol.* **1969**, *101*, 1033-1041.
46. Lewinsohn, E.; Gijzen, M.; Muzika, R. M.; Barton, K.; Croteau, R. *Plant Physiol.* **1992**, in press.
47. Person, H. L. *J. For.* **1931**, *29*, 696-699.

48. Christiansen, E.; Horntvedt, R. *Z. Ang. Entomol.* **1983**, *96*, 110-118.
49. Brand, J. M.; Schultz, J.; Barras, S. J.; Edson, L. J.; Payne, T. L.; Hedden, R. L. *J. Chem. Ecol.* **1977**, *6*, 657-666.
50. Borden, J. H. In *Bark Beetles in North American Conifers: A System for the Study of Evolutionary Biology*; Mitton, J. B.; Sturgeon, K. B., Eds.; University of Texas Press: Austin, TX, 1982; pp 74-139.
51. Brand, J. M.; Bracke, J. W.; Britton, L. N.; Markovetz, A. J.; Barras, S. J. *J. Chem. Ecol.* **1976**, *2*, 195-199.
52. Vité, J. P.; Francke, W. *Naturwissenschaften* **1976**, *63*, 550-555.
53. Wood, D. L. *Ann. Rev. Entomol.* **1982**, *27*, 411-446.
54. Silverstein, R. M.; Rodin, J. O.; Wood, D. L. *Science* **1966**, *154*, 509-510.
55. Birch, M. C.; Light, D. M.; Wood, D. L.; Browne, L. E.; Silverstein, R. M.; Bergot, B. J.; Ohloff, G.; West, J. R.; Young, J. C. *J. Chem. Ecol.* **1980**, *6*, 703-717.
56. Mustaparta, H.; Angst, M. E.; Lanier, G. N. *J. Chem. Ecol.* **1980**, *9*, 689-701.
57. Lanier, G. N.; Classon, A.; Stewart, T.; Piston, J. J.; Silverstein, R. M. *J. Chem. Ecol.* **1980**, *9*, 677-688.
58. Raffa, K. F.; Klepzig, K. D. *Oecologia* **1989**, *80*, 566-569.
59. Renwick, J. A. A.; Hughes, P. R.; Krull, I. S. *Science* **1976**, *191*, 199-201.
60. Francke, W.; Vité, J. P. *Z. Ang. Entomol.* **1983**, *96*, 146-156.
61. Pitman, G. B.; Vité, J. P.; Kinzer, G. W.; Fentiman, A. F. Jr. *Nature* **1968**, *218*, 168-169.
62. Hughes, P. R. *Z. Ang. Entomol* **1973**, *73*, 294-312.
63. Pitman, G. B. *J. Econ. Entomol.* **1971**, *64*, 426-430.
64. Hunt, D. W. A.; Borden, J. H.; Lindgren, B. S.; Gries, G. *Can. J. For. Res.* **1989**, *19*, 1275-1282.
65. Mori, K. *Tetrahedron* **1989**, *45*, 3233-3298.
66. Leufven, A.; Nehls, L. *Microb. Ecol.* **1986**, *12*, 237-243.
67. Leufven, A.; Birgersson, G. *Can J. Bot.* **1987**, *65*, 1038-1044.
68. Hunt, D. W. A.; Borden, J. H. *J. Chem. Ecol.* **1990**, *16*, 1385-1397.
69. Byers, J. A.; Birgersson, G. *Naturwissenschaften* **1990**, *77*, 385-387.
70. Hendry, L. B.; Piatek, B.; Browne, L. E.; Wood, D. L.; Byers, J. A.; Fish, R. H.; Hicks, R. A. *Nature* **1980**, *284*, 485.
71. Dahlsten, D. L. In *Bark Beetles in North American Conifers: A System for the Study of Evolutionary Biology*; Mitton, J. B.; Sturgeon, K. B., Eds.; University of Texas Press: Austin, TX, 1982; pp 140-182.
72. Wainhouse, D.; Wyatt, T.; Phillips, A.; Kelly, D. R.; Barghian, M.; Beech-Garwood, P.; Cross, D.; Howell, R. S. *Chemoecology.* **1989**, *2*, 55-63.
73. Baisier, M.; Grégoire, J.-C.; Delinte, K.; Bonnard, O. In *Mechanisms of Woody Plant Defenses Against Insects. Search for Pattern*; Mattson, W. J.; Levieux, J.; Bernard-Dagan, C., Eds.; Springer-Verlag: New York, NY, 1988; pp 359-368.
74. Everaerts, C.; Grégoire, J.-C.; Merlin, J. In *Mechanisms of Woody Plant Defenses Against Insects. Search for Pattern*; Mattson, W. J.; Levieux, J.; Bernard-Dagan, C., Eds.; Springer-Verlag: New York, NY, 1988; pp 335-344.
75. Gershenzon, J.; Croteau, R. In *Biochemistry of the Mevalonic Acid Pathway to Terpenoids*; Towers, G. H. N.; Stafford, H. A., Eds.; Recent Advances in Phytochemistry; Plenum Press: New York, NY; 1990, Vol. 24; pp 99-160.

76. Gershenzon, J.; Croteau, R. In *Herbivores: Their Interactions with Secondary Metabolites*; Rosenthal, G. A.; Berenbaum, M., Eds.; Academic Press: New York, NY, 1991; pp 165-219.
77. Croteau, R. *Chem. Rev.* **1987**, *87*, 929-954.
78. Wagschal, K.; Savage, T. J.; Croteau, R. *Tetrahedron* **1991**, *47*, 5933-5944.
79. Hanover, J. W. *Heredity* **1966**, *21*, 73-84.
80. Hanover, J. W. *Heredity* **1971**, *27*, 237-245.
81. Squillace, A. E.; Wells, O. O.; Rockwood, D. L. *Silv. Genet.* **1980**, *29*, 141-151.
82. Savage, T.; Croteau, R. *Naval Stores Rev.* **1991**, *101*, 6-11.
83. Gijzen, M.; Lewinsohn, E.; Croteau, R. *Arch. Biochem. Biophys.* **1991**, *289*, 267-273.
84. Lewinsohn, E.; Gijzen M.; Croteau, R. In *Regulation of Isopentenoid Metabolism*; Nes, W. D.; Parish, E. J.; Trzaskos, J. M., Eds.; ACS Symposium Series, American Chemical Society; Washington, DC; 1992, in press.
85. Lewinsohn, E.; Gijzen, M.; Croteau, R. *Arch. Biochem. Biophys.* **1992**, *293*, 167-173.
86. Gijzen, M.; Lewinsohn, E.; Croteau, R. *Arch. Biochem. Biophys.* **1992**, *294*, 670-674.
87. Huang, Y.; Diner, A. M.; Karnosky, D. F. *In Vitro Cell. Dev. Biol.* **1991**, *91*, 201-207.
88. van Wordragen, M. F.; Dons, H. J. M. *Plant Mol. Biol. Rep.* **1992**, *10*, 12-36.

RECEIVED August 19, 1992

Chapter 3

Volatile Components of Tomato Fruit and Plant Parts
Relationship and Biogenesis

Ron G. Buttery and Louisa C. Ling

Western Regional Research Center, Agricultural Research Service, U.S. Department of Agriculture, Albany, CA 94710

A comparison is made of the volatile component concentrations in the macerated (blended) forms of tomato fruit and leaf and to a lesser extent in macerated flower, calyx and stem. These studies show that some processes such as the formation of C6 lipid derived volatiles are similar between fruit and leaf. Formation of higher carbon number lipid derived volatiles (e.g. 4,5-epoxy-(E)-2-decenal), however, only occurs in the fruit. Other processes such as the formation of carotenoid related (e.g. 6-methyl-5-hepten-2-one) and amino acid related (e.g. nitro, nitrile and thiazole) volatiles seem to be largely confined to the fruit. Terpenoid hydrocarbons occur in the leaf but not the fruit, except for (-)-alpha-copaene in the mature green fruit. Lipid oxidation processes occur in the first few minutes after maceration but glycoside hydrolysis and amino acid derived volatile biogenesis seems to occur during the ripening process in the intact tomato.

The authors have been carrying out a continuing study of the volatile aroma and flavor components of tomatoes. The control of tomato flavor, in both the agricultural production of tomatoes and their processing, could benefit from basic knowledge of the biogenesis or chemogenesis of the flavor components. This information is also important to any genetic approaches in flavor control.

The leaves and other plant parts have aromas which bear a resemblance to the aroma of the fruit. Using special tissue culture techniques it has been found that some plant tissue can be transformed into ripe fruit like tissue (1). It seemed then that knowledge of the volatiles of different plant parts may give us some clues to biogenesis in the fruit. The authors have therefore made a study of the volatiles of tomato leaves and other plant parts and compared the volatiles to those of the fruit.

Division of Flavor Components into Groups. To simplify the study it is useful to divide the tomato (fruit and plant) volatiles into several groups. These are listed in Table I. We will discuss each group separately.

In general, unless otherwise mentioned, the method of isolation of volatiles and their capillary GLC- MS analysis is similar to that described previously by the authors. This involved first blending of the fruit or plant material with a small amount of water, holding of the blended material for 3 minutes for enzyme generation of volatiles, then addition of excess saturated $CaCl_2$ solution to halt further enzyme action. After addition of internal standards the volatiles were swept from the vigorously stirred mixture to a large (10g) Tenax trap with a flow (3L/min.) of purified air for 1 hour. Extraction of the trap with diethyl ether and distillation removal of most solvent gave the concentrate for GLC-MS.

Lipid Derived Volatiles.

Volatiles thought to be derived from lipids, principally unsaturated fatty acids, are listed in Table II.

Also listed in Table II is a comparison of the concentrations found for these compounds in the blended fruit and leaf. Most of these compounds had been identified in previous studies of tomato volatiles (2). Two compounds, however, were only recently identified in tomatoes for the first time by the authors. These are 4,5-epoxy-(E)-2-heptenal (2 isomers) and 4,5-epoxy-(E)-2-decenal (2 isomers). Their mass spectra and GLC retention times are consistent with those of authentic samples (C7 epoxide major MS ion at 68,39,29,55,81,97,110 and GLC K.I. DB-1 1036: C10 epoxide major MS ions at 68,39,29,55,81,95,139 and K.I. 1340). An unidentified compound (apparently $C_6H_8O_2$), occurring at ca. 1 ppm concentration in the leaves and in a lower level in the fruit, had a mass spectrum similar to that reported for 2-hexen-4-olide which had been identified previously in raspberry (3), asparagus (4) and bread volatiles (5), however, the GLC retention index of the tomato unknown was somewhat different from that reported (5).
Epoxides of the type identified in tomatoes were first discovered by Swoboda and Peers (6) in copper catalyzed vegetable oil oxidation and more recently found by Grosch and coworkers (cf.7) in soybean oil and related products. They are fairly unstable compounds. The inertness of fused silica capillary columns has made their analysis more feasible. The 4,5-epoxy-(E)-2-decenal occurs in the blended tomato fruit up to 100 ppb but we have been unable to find it or the C_7 compound in the blended leaves. The 2 main isomers (E,Z)- and (E,E)- of 2,4-decadienal occur at less than 1/3 the concentration of the C_{10} epoxide so that it seems that this compound probably results from some oxidized lipid precursor rather than be formed by oxidation of the dienal.

The epoxides are difficult to isolate in their pure forms to determine their odor thresholds in water solution. However, Grosch and coworkers (7) have used a method (called AEDA) of threshold determination in the carrier gas effluent of the fused silica capillary columns and found the C_{10} epoxide aldehyde to be a moderately potent odorant.

Table I. Major groups of volatiles in fresh tomatoes and leaves

I.	Lipid derived.
II.	Carotenoid related.
III.	Amino acid related.
IV.	Terpenoids (C_{10} and C_{15}).
V.	Lignin related and other miscellaneous.

Table II. Lipid derived volatiles identified in fresh tomato fruit and fresh plant leaves blended and held 3 min. before enzyme deactivation

Compound[a]	Conc. in ppb[b]	
	Fruit	Leaves
1-Penten-3-one	450	2000
1-Penten-3-ol	100	2400
Pentanal	5	<5
(Z)-and (E)-Pentenal	100[c]	100[c]
Pentanol	30	430
(Z)-2-Pentenol	40	1300
(Z)-3-Hexenal	15000	220000
Hexanal	2000	3900
(E)-2-Hexenal	470	17000
(Z)-3-Hexenol	120	10000
Hexanol	4	500
(E,Z)-and (E,E)-2,4-Hexadienal	10	180
Unknown $C_6H_8O_2$	74	870
(E)-2-Heptenal	40	d
1-Octen-3-one	<5	d
(E,Z)- and (E,E)-2,4-Heptadienal	5[c]	d
4,5-epoxy-(E)-2-Heptenal (2 isomers)	4	d
(E)-2-Octenal	15	d
(E,Z)-and (E,E)-2,4-Decadienal	10	d
4,5-epoxy-(E)-2-Decenal (2 isomers)	30	d

[a] Mass spectra and GLC retention index consistent with that of authentic sample.
[b] ppb = parts of compound per billion (10^9) parts of tomato or leaf.
[c] Not completely resolved from other peaks: not possible to accurately measure concentration.
[d] Not detected by MS. If present < 5 ppb

C_6 (Green Odor) Compounds.

The C_6 compounds are by far the predominant compounds in both the blended leaves and fruit. (Z)-3-hexenal was found to occur at 220 ppm in the blended leaves which is 10 times higher than the concentration in the fruit. An earlier study by the authors (8) had indicated a much lower concentration of (Z)-3-hexenal (23 ppm) but a 270 ppm concentration of (E)-2-hexenal. In that earlier study the (Z)-3-hexenal was apparently largely isomerized to the (E)-2-hexenal during the isolation or analysis. This isomerization can occur very readily particularly at low pH and by various subtle catalysts.

The rates of formation of some of the C_6 compounds in the leaves were studied by holding samples of blended leaves (magnetically stirred) in a flask at 25°C and removing 0.1 ml samples of the vapor for GLC analysis (through a small hole in the flask covered with Teflon film) each time period. A plot of the data obtained for (Z)-3-hexenal is shown in Figure 1. It can be seen that the (Z)-3-hexenal increases very rapidly and reaches a maximum in about 2-5 minutes. There is then a gradual loss due to isomerization to (E)-2-hexenal and reductase conversion to (Z)-3-hexenol. Other chemical reactions probably also go on including acetal type condensation with the sugars. We also carried out this type of study with (Z)-3-hexenal in tomato fruit which showed a very similar rapid rise with a maximum at about 3 minutes and gradual loss, although, at a ca. 10 times lower concentration than the leaves.

In our previous study of tomato leaves (8), the evidence had indicated that the (Z)-3-hexenal content (found then) was sufficiently high to be responsible for the characteristic green, tomato fruit like aroma given off by tomato leaves when they are handled etc. The much higher concentration found in the present study would tend to give stronger support to this idea.

To find the concentration for zero time, or what was presumably present in the intact tomato leaves, the authors blended the leaves in saturated $CaCl_2$ solution. This showed 10 ppm (Z)-3-hexenal. However, it was thought that some enzyme action may have gone on in the maceration of the leaves before the $CaCl_2$ had time to penetrate and be effective. Another method that was used was to heat the leaves very rapidly to 100°C (20 seconds) in a microwave oven to deactivate the enzymes. The leaves were then immediately dropped into an excess of ice cold saturated $CaCl_2$, the mixture blended, and the volatiles isolated by dynamic headspace sampling in the usual way. This showed a much lower concentration of (Z)-3-hexenal at 0.01 ppm (10 ppb).

Both of the methods of deactivation of the enzymes at zero time ($CaCl_2$ and microwave) were used to study whether particular volatiles were present in the intact fruit or whether they were only formed on maceration which gives some clue to their biogenesis. Both methods have flaws. The $CaCl_2$ method has the disadvantage that there is a slight delay before the $CaCl_2$ can penetrate and be effective. The high temperature of the microwave method, although brief, could possibly result in some volatization and also some chemical reactions such as acetal formation between aldehydes and sugars.

With tomato fruit a study was also made of the concentration of (Z)-3-hexenal at the various ripening stages. This comparison with the blended fruit showed 0.1 ppm concentration in the mature green tomato, 0.3 ppm concentration in the breaker stage and 15 ppm concentration, in the table ripe tomato.

The general biosynthetic pathway for C_6 compounds from unsaturated fatty acids has, of course, been thoroughly studied by Hatanaka and coworkers (9) for tea leaves and for tomato fruit by Galliard and coworkers (10) and Stone and coworkers (11).

Carotenoid Related Volatiles.

These are listed in Table III. These compounds (mostly norterpenoids) have only been identified in the fruit. The authors have been unable to detect these in the normal green leaves, stems, flowers or calyx. There are two main types of compounds in this group, linear and cyclic. A comparison is shown also for the concentrations found in the intact tomato and those found in macerated tomato held 3 minutes before enzyme deactivation. One unexpected finding is that there seems to be a different principal mechanism for the formation of geranylacetone than there is for 6-methyl-5-hepten-2-one. They are related compounds, the geranyl-acetone being one C5 isoprene type unit greater than the methylheptenone. The geranylacetone is increased 10-fold in the macerated tomato whereas the methyl-heptenone is only increased 2-fold. The authors had also previously noticed in their studies on processed tomato that the concentration of methylheptenone increased on heating whereas this was not the case with geranylacetone. The methylheptenone was also found amongst the products of pH 4 thermal hydrolysis of tomato glycosides (12) and it seems that one pathway to methylheptenone is through the glycosides. The large increase in geranylacetone on maceration, on the other hand and its absence in the glycoside hydrolysis products, indicates that its principal mode of formation is by an oxidative process possibly coupled with the lipid oxidation.

The (E)-6-methyl-3,5-heptadien-2-one was also found amongst the products of hydrolysis of the tomato glycoside fraction and thus this seems to be a likely pathway for its formation.

The very low (but important) concentration of beta-ionone in tomatoes makes it difficult to obtain reliable quantitative data on it, but, like geranylacetone it seems also to be formed mainly by an oxidative mechanism. It was not detected amongst the glycoside hydrolysis products.

The compound beta-damascenone has been shown to be produced in fruits (13) from hydrolysis of glycosides via an intermediate acetylenic compound megastigm-5-en-7-yne-3,9-diol and this seems to be its main mode of formation in tomatoes. The authors had found beta-damascenone as a major component of the volatiles from thermal pH 4 hydrolysis of tomato glycosides. Recently Marlatt and coworkers (14) reported the isolation of the above acetylene compound from enzyme hydrolysis of tomato glycosides, thus confirming that this acetylene compound (as with fruits) is also a major intermediate in the route to beta-damascenone in tomatoes.

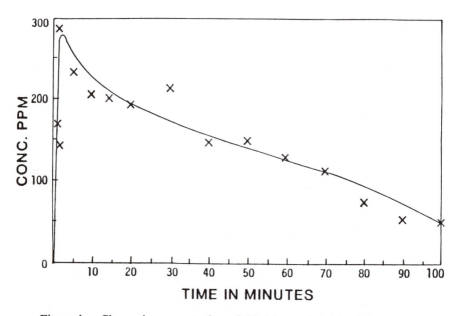

Figure 1.   Change in concentration of (Z)-3-hexenal in blended tomato leaves with time as found by direct headspace sampling.

Table III.   Carotenoid related fresh tomato volatiles [a]

Compound	Conc.(ppb) "intact" [b]	Conc.(ppb) macerated
Open Chain		
6-Methyl-5-hepten-2-one	100	210
6-Methyl-5-hepten-2-ol	8	8
Geranylacetone	20	330
Pseudoionone	11	6
Cyclic		
2,2,6-Trimethylcyclohexanone	<5	<5
beta-Cyclocitral	3	5
beta-Damascenone	<5	<5
beta-Ionone	11	18
epoxy-beta-Ionone	<5	<5

[a] No significant amounts (> 5 ppb) of this group have been found in the leaves, flowers   or stems.

[b] Enzymes deactivated before or during blending as described in text.

Amino Acid Related Volatiles.

Fresh tomato fruit volatiles related to amino acids are listed in Table IV. Again, except for 2-phenylethanol and 2- and 3-methylbutanols, these compounds were not detectable in the leaves, stems, flowers or calyx. Concentrations found for the "intact" tomato isolation were compared to those for the 3 minute held macerated tomato. It was was found that there is very little difference for most of these compounds. They thus seem to be formed in the intact tomato during the ripening process. We have also looked at the concentrations of these compounds in tomato fruit at the different stages of ripening in the mature green, breaker (slightly pink) and table ripe stages. The main formation of most of these compounds seems to occur between the breaker and table ripe stages.

Nitro Compounds and Nitriles. Volatile nitro compounds are unusual in foods and don't seem to have been reported to occur in any other food than tomato. Nitro compounds of similar types, however, had been identified in night blooming flowers by Kaiser (*15*). Analysis of tomato flowers did not show any nitro compounds, eliminating the flowers as a possible source of the compounds in tomato fruit. Kaiser (*15*) had proposed that the biochemical pathway for the formation of the nitro compounds in flowers (Figure 2) is a variation of an earlier one by Cohn (*16*) for the biosynthesis of cyanogenic glycosides. As pointed out by Kaiser, the pathway leads to the formation of both nitro compound and nitrile. Both predicted nitro and nitrile forms do occur in tomato (Table V) for the leucine and phenylalanine derived compounds. We are well aware of the spectral and GLC retention data on synthetic forms of the intermediate oximes but have been unable to detect the natural compounds in the fresh tomato. The 3-methyl-butanal oxime was, however, reported to be identified in cooked tomato products by one group of researchers (*17*). No spectral or GLC retention data were given, although the research team involved has a good reputation for accuracy.

Alkylthiazoles. Alkylthiazoles are well known in cooked products, but tomatoes are one of the few fresh foods where an alkylthiazole has been found. Although some isolated claims have been made for a number of alkylthiazoles in fresh tomato fruit, 2-isobutylthiazole is the only one that has been identified with certainty by many different researchers. 2-Isobutylthiazole occurs in the ripe fruit at ca. 100-300 ppb but we have been unable to detect any in the leaves or other parts of the tomato plant.

There have been some hypotheses regarding how 2-isobutylthiazole is formed in the tomato but little real evidence. The hypothesis put forward by Schutte (*18*) seems the most plausible. This scheme suggests first the formation of 3-methylbutanal in the fresh tomato (of which there is a reasonable concentration) from breakdown of leucine. This then condenses with cystamine (possibly formed from cysteine) to give a thiazolidine (a cyclic N, S analog of an acetal) which could possibly oxidize enzymatically to the thiazole. Evidence linking leucine to 2-isobutylthiazole using isotope methods was reported by Stone and coworkers (*19*). This type of condensation of cystamine with aldehydes goes very readily in good yield and has been used by Shibamoto and coworkers (*20*) to make stable derivatives of aldehydes for their quantitative analysis in food systems.

Table IV. Volatiles in fresh tomatoes (blended) related to amino acids[a]

Amino Acid	Compound	Conc. ppb Fruit
Alanine	Acetaldehyde	800
Valine	1-Nitro-2-methylpropane	<5
Leucine	3-Methylbutanal	27-65
	3-Methylbutylnitrile (isobutyl cyanide)	13-42
	3-Methylbutanol	150-380
	1-Nitro-3-methylbutane	59-300
	3-Methylbutyric acid	200
	2-Isobutylthiazole	36-110
Isoleucine	2-Methylbutanol	100
	2-Methylbutyric acid	5
Phenyl-alanine	Phenylacetaldehyde	15-18
	2-Phenylethanol	1000
	1-Nitro-2-phenylethane	17-54
	Phenylacetonitrile	3-8

[a] Of this group only 2-and 3-methylbutanol and 2-phenylethanol occur in the leaves in relatively small concentrations.

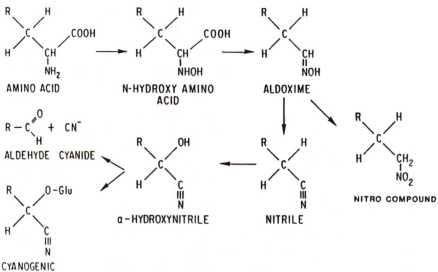

Figure 2. Biosynthetic pathway to nitro compounds in tomatoes is possibly a variation of the known pathway to cyanogenic glycosides (16) as suggested by Kaiser (15) for night blooming flowers.

Besides 3-methylbutanal, fresh tomatoes also have moderate concentrations of phenylacetaldehyde and one might expect that this would also react in the same way to form the corresponding thiazole. We synthesized the expected 2-phenylmethylthiazole to obtain its GLC retention time and mass spectrum. In analysis of tomato volatiles, however, we could find no evidence for this compound. The enzyme systems involved must be specific for producing only the 2-isobutylthiazole.

Amino Acid Derived Aldehydes. The aldehydes 2-and 3-methylbutanals and phenylacetaldehyde are well known Strecker degradation products in cooked foods from amino acid: sugar browning type reactions. But they are also present in fresh unheated tomato at a reasonable concentration and can not arise there through browning type reactions. There is a question then of the mechanism of formation of these compounds in fresh tomato. One possible pathway is by enzymatic oxidation of the corresponding alcohols released by enzymatic hydrolysis of glycosides occurring during ripening. Another possibility is an enzymatic version of the so called Nef reaction which involves conversion of the nitro compounds to the corresponding aldehyde by first making a solution of the nitro compound alkaline and then acidifying it. Very mild conditions (pH 10 and pH 4.5) were enough to convert the 1-nitro-2-phenylethane to phenylacetaldehyde in our laboratory and it is reasonable that an enzyme system in tomato might also bring about a change of this type.

Terpenoids (C_{10} and C_{15}).

In contrast to the situation with amino acid derived compounds, the ripe fruit contains no terpenoid hydrocarbons, although they are major components of the leaves and other parts of the tomato plant both in the intact and whole blended forms. One exception that the authors recently found, though, is that (-)-alpha-copaene occurs in the blended mature green tomatoes (capillary GLC and MS consistent with that of an authentic sample). The authors were able to establish the direction of optical rotation from the use of a permethylated beta-cyclodextrin fused silica capillary column (cf.*21*). The major terpenoids found in the blended leaves and fruit are listed in Table VI. The hydrocarbons also occur in the blended stem, flower and calyx in lesser concentrations. These compounds and others had been previously identified by the authors (*8*) and others (*22*) in the whole tomato leaf. Some of the terpenoids seem to be lost as a result of the blending, possibly through lipid coupled oxidative degradation. The 2-carene occurring in tomatoes was found to have the same retention time as (+)-2-carene on a permethylated beta-cyclodextrin fused silica capillary GLC column. The biosynthetic pathways leading to terpenoids have been well established by other researchers.

The only (C_{10}) oxygenated terpenoids that the authors have found in the blended fresh fruit are linalool, neral and geranial. Linalool has commonly been found in cooked fruits as a product of glycoside hydrolysis and we (*12*) also showed that it occurs in the products of thermal acidic hydrolysis of tomato

Table V. Nitro and nitrile compounds in fresh ripe tomatoes

Compound	Conc. ppb	
	"intact"[a]	blended
1-Nitro-2-methylpropane	<5	<5
3-Methylbutyronitrile	75	65
1-Nitro-3-methylbutane	300	300
Phenylacetonitrile	5	7
1-Nitro-2-phenylethane	51	27

[a] Enzymes deactivated before or during blending as described in text.

glycosides. It has also been found amongst the products of enzymatic hydrolysis of tomato glycosides (14).

Lignin Related and Other Miscellaneous Volatiles.

These are listed in Table VII together with their concentrations in both blended fruit and leaves. All compounds occur in all plant parts. Although in the present study we found a lower amount in the blended leaves, we previously found 1 ppm eugenol by direct solvent extraction of the whole leaves (8). It is a potent odorant with a threshold of 6 ppb in water solution and must contribute consid-erably to the odor of the leaves. Being a phenol it has strong adsorption properties and a low volatility in water.   This makes it difficult to get a good recovery in quantitative analyses.

Benzaldehyde and benzoic acid very probably result from glycoside hydrolysis in the fresh tomato. We (12) had found benzaldehyde and others (14) had found benzoic acid in the products of enzymatic hydrolysis of tomato glycosides. The authors (8) had also found eugenol and guaiacol in the products of pH 4 heated tomato glycoside hydrolysis. Methyl salicylate is one of the most quantitatively variable compounds in both leaves and fruit and the authors have been unable to correlate this variability with any difference in tomato lines, maturity or growing conditions.

Comparison of Leaf, Calyx, Flower and Stem Volatiles.

GC-MS studies of the volatiles from the 4 different plant parts (blended) showed them to be very similar qualitatively.   However, they did show quantitative differences as can be seen for a few of the components in Table VIII. The leaves showed the highest concentration of (Z)-3-hexenal and the stem the lowest. The authors expected the flowers to yield some more unusual compounds but this was not the case and only compounds already identified in the leaf were detected.

Table VI. Volatile C_{10} and C_{15} terpenoids in blended tomato leaves and fruit

Compound	Conc. ppb leaves	Conc.ppb fruit
alpha-Pinene	100	<1
(+)-2-Carene	1700	<1
Limonene	1000	<1
beta-Phellandrene	8000	<1
Linalool	<10	2
Neral	<5	2
Geranial	<5	12
(-)-alpha-Copaene	<5	10 (green)[a]
Caryophyllene	350	<1
Humulene	250	<1
Caryophyllene epoxide	64	<1

[a] Only found in mature green fruit.

Table VII. Lignin related and other miscellaneous volatiles in blended tomato fruit and leaf

Compound	Fruit ppb	Leaves ppb
Benzaldehyde	31	30
Benzyl alcohol	20	30
2-Methoxyphenol (guaiacol)	15	10
Methyl salicylate	48	100-600
Eugenol	100	100-500

Table VIII. Comparison of concentrations found for volatiles in blended fresh plant parts. All were found similar qualitatively

Compound	Conc. ppm			
	Leaf	Calyx	Flower	Stem
(Z)-3-Hexenal	220	43	4	2
(+)-2-Carene	2	0.3	5	0.4
Methyl salicylate	0.6	0.4	6	1
Caryophyllene	0.4	0.03	3	0.02

LITERATURE CITED

1. Ishida, B. K. *The Plant Cells*, **1991**, *3*, 219-213.
2. Petro-Turza, M. *Food Reviews International*, **1986-87**, *2*, 309-351.
3. Winter, M.; Enggist, P. *Helv Chim. Acta.*, **1971**, *54*, 1891-1898.
4. Tressl, R.; Bahri, D.; Holzer, M.; Kossa, T. J. *Agric. Food Chem.* **1977**, *25*, 459-463.
5. Schieberle, P.; Grosch, W. Z. *Lebensm.-Unters. Forsch.*, **1983**, *177*, 173-180.
6. Swoboda, P. A. T.; Peers, K. E. *J. Fd. Agric.*, **1978**, *29*, 803-807.
7. Guth, H.; Grosch, W. *Lebensm.-Wissen. u.-Technol.*, **1990**, *23*, 513-522.
8. Buttery, R. G.; Ling, L. C.; Light, D. M. *J. Agric. Food Chem.*, **1987**, *35*, 1039-1042.
9. Hatanaka, A.; Kajiwara, T.; Sekiya, T. *Lipids*, **1987**, *44*, 341-361.
10. Galliard, T.; Matthew, J. A.; Wright, A.; Fishwick, M. *J. Phytochem.* **1976**, *15*, 1647-1651. *J. Sci. Fd. Agric.* **1977**, *28*, 863-868.
11. Stone, E. J.; Hall, R. M.; Kazeniac, S. J. *J. Food Sci.*, **1975**, *40*, 1138-1141.
12. Buttery, R. G.; Takeoka, G.; Teranishi, R.; Ling, L. C. *J. Agric. Food Chem.*, **1990**, *38*, 2050-2053.
13. Sefton, M. A.; Skouroumounis, G. K.; Massy-Westropp, R. A.; Williams, P. J. *Aust. J. Chem.*, **1989**, *42*, 2071-2084.
14. Marlatt, C.; Ho, C.-T.; Chien, M. *J. Agric. Food Chem.*, **1992**, *40*, 249-252.
15. Kaiser, R. Paper presented at EUCHEM 1987, "Semiochemicals in the Plant and Animal Kingdoms", Anger, France, Oct. **1987**.
16. Cohn, E. E. *Naturwissenschaften*, **1979**, *66*, 28-34.
17. Wobben, H. J.; de Valois, P. J.; ter Heid, R.; Boelens, H.; Timmer, R. *Proc. IV Int. Congress Food Sci. Technol.*, **1974**, p 22-24.
18. Schutte, L. *CRC Critical Reviews in Food Tech.*, **1974**, *4*, 457-505.
19. Stone, E. J.; Hall, R. M.; Kazeniac, S. J. **1971**, Paper presented at 162nd National Meeting, American Chemical Society, Washington, D.C.
20. Sakaguchi, M.; Shibamoto, T. *J. Agric. Food Chem.*, **1978**, *26*, 1179-1183.
21. Takeoka, G.; Flath, R. A.; Mon, T. R.; Buttery, R. G.; Teranishi, R. *J. High Resol. Chrom.*, **1990**, *13*, 202-206.
22. Anderson, B. A.; Holman, R. T.; Lungren, L.; Stenhagen, G. *J. Agric. Food Chem.*, **1990**, *35*, 1039-1042.

RECEIVED December 14, 1992

Chapter 4

Semio Activity of Flavor and Fragrance Molecules on Various Insect Species

Braja D. Mookherjee[1], Richard A. Wilson[1], Kenneth R. Schrankel[1], Ira Katz[1], and Jerry F. Butler[2]

[1]International Flavors and Fragrances, Inc., 1515 Highway 36, Union Beach, NJ 07735
[2]Institute of Food and Agricultural Sciences, Department of Entomology and Nematology, University of Florida, Building 970, IFAS 0740, Gainesville, FL 32611

An exhaustive screening program on more than 2000 flavor and fragrance molecules and natural products was performed using a newly designed multiport olfactometer and was supported by field trials. The attractancy or repellancy of a number of these materials with respect to several insect species including housefly (*Musca domestica*), mosquitoes (*Aedes egyptae*), sandflies (*Psychodidae*), beetles (*Coleoptera)*, and stored product moths will be discussed.

Semio chemicals can be broadly classified into two categories: pheromones, which are molecules or combinations of molecules excreted by living members of the animal kingdom, especially insects, for the primary purpose of eliciting a behavior modification in a member of its own species; and molecules and combinations of molecules which function, for the most part incidentally, to attract or repel an animal organism. A search of the literature shows a plethora of publications dealing with pheromones but there has been very little published regarding the effect of flavor and/or fragrance molecules on insects. It is reported that there is some evidence that a few fragrance chemicals have shown semio activity. For example, the substances shown in Table I have been reported to attract various insect species.

Table II shows some substances which are reported to repel insects.

There have also been reports of mixtures of chemicals which show attractant semio activity (see Table III).

In addition, the mixtures of compounds shown in Table IV have been reported to act as repellants.

However, to date there has been no systematic investigation of the semio activity of common flavor and fragrance molecules. We, therefore, undertook the task of screening a large sampling of these materials for attractancy or repellancy toward some of the most common household insect pests. For this

0097–6156/93/0525–0035$06.00/0
© 1993 American Chemical Society

TABLE I.

Fragrance Chemicals and Natural Products Reported to Attract Insects

Fragrance Material	Attracts
trans-6-Nonenol	Female Melon Fruitfly (1)
Indian Calamus Root	Female Mediterranean Fruitfly,
Oil (Acorus calamus)	Female Melon Flies, Male and Female Oriental Fruit flies (2)
Ethanol (ex Fermented Molasses or Sucrose)	Little Houseflies (3)
Methyl Propyl Disulfide, cis and trans Propenyl Propyl Disulfide (Components of onion, Allium cepa)	Hylemya antigua (4)
Phenylacetaldehyde	Moths (many species) (5)
cis-6-Nonenal (ex Melon)	Melon Fruitfly (6)
trans-2-Hexenal (ex Oak Leaves)	Female Moths (7)
alpha Farnesene (ex Apples)	Codling Moth (7)
Caryophyllene	Cotton Insect (7)
Methyl Eugenol	Oriental Fruitfly (7)
Limonene	Fruitflies (7)

purpose, we selected two primary insects. These are the common housefly (*Musca domestica*), shown in Figure 1, and the mosquito (*Aedes egyptae*), which is shown in Figure 2.

Both of these insects are also major vectors for disease, therefore, methods to control their population in the environment are obviously of particular interest. In addition, selected chemicals were screened for activity toward stored clothes moths such as those shown in Figure 3.

TABLE II.

Fragrance Materials Reported to Repel Insects

Fragrance Material	Repels
Japanese Vetiver Oil (Carbonyls)	Cockroaches and Flies (8)
Japanese Mint and Scotch Spearmint Components	Cockroaches (9)
(-) Limonene, (-) Menthol, (-) Menthone, (-) Carvone, (-) Pulegone	
(Their enantiomers and racemic mixtures had very low activity.)	
Cucumber Skin (*trans*-2-Nonenal)	Cockroaches (10)
Bay Leaves (1,8-Cineole)	Cockroaches (11)
Citronella Oil (Citronellol, Citronellal)	Mosquitoes (12)

Experimental

The olfactometer (19) which was employed in the laboratory screening was designed and built at the University of Florida. It is shown in Figure 4.

It is a pie-type with 4 to 10 choice ports, normally run with 10 choice treatments. Treatments are made to the airstream and to surface "skins" representing an artificial host which are placed within the perimeter of the test chamber. Insects are placed in the center of the test chamber.   They move to the perimeter and to the artificial hosts down the airstream in a time series fashion depending upon the attractancy or repellancy of the treated air and "skin" surface. Visual and electronic counts are made on the activity of the introduced sample at timed intervals of 10 minutes to 24 hours. Activities at the choice points are compared to the activities of standard attractants and repellants which are tested at the same time.

Selected materials which have been identified as semio-active in the laboratory are subjected to field tests during periods of high activity for selected insects. Various types of field traps were employed depending on the type of insect being studied.   One type of field olfactometer employed (patent pending - IFF) is shown in Figure 5. By this method, various insects such as beetles and sandflies, among others, are attracted to varying degrees to selected test chemicals.

TABLE III.

Chemical Mixtures Reported to Attract Insects

Mixture	Attracts
trans-2-Hexenal, *alpha* Terpineol, Benzyl Alcohol, Linalool, and 4-Terpinenol	Soldierbugs (*13*)
trans Methyl Jasmonate, *epi* Methyl Jasmonate, Ethyl *trans* Cinnamate, R-(-) Mellein	Female Oriental Fruit Moths (*14*)

		Male and Female
sec. Butyl Alcohol	(10.6%)	Screwworms (*15*)
*iso*Butyl Alcohol	(10.3%)	
Butyric Acid	(12.5%)	
Acetic Acid	(13.7%)	
Dimethyl Disulfide	(12.0%)	
Phenol	(11.9%)	
para Cresol	(11.5%)	
Indole	(2.6%)	
Benzoic Acid	(2.6%)	

(This mixture of compounds was considered likely to be found in decomposing animal or bacterial products.)

Mixture	Attracts
Phenyl Ethyl Propionate and Eugenol	Both Sexes of Japanese Beetle (*16*)
2,4-Hexadienyl Butyrate, and Heptyl Butyrate or Octyl Butyrate	Yellow Jacket Hornets (*16*)
Methyl *iso*Eugenol, Methyl Eugenol and Veratric Acid	Fruitflies (*17*)

TABLE IV.

Chemical Mixtures Reported to Repel Insects

Mixture	Repels
Short-chain Ketones, Formaldehyde, and Propionaldehyde	Tsetse Flies (*18*)
Long-chain Ketones, Heptaldehyde, and Caproic Acid	Tsetse Flies (*18*)

Figure 1.   Houseflies (*Musca domestica*)

Figure 2.   Mosquitoes (*Aedes egyptae*)

Figure 3.   Clothes Moths

Figure 4.   Olfactometer

Figure 5.   Field Trap

Results and Discussion

Using the olfactometer, we have tested over 2000 materials which have included common flavor and fragrance synthetic chemicals as well as flower, fruit, plant, and animal essential oils and extracts. Many of these were shown to be neither attractants nor repellants. For example, the highly odorous and extensively used nature-identical flavor and fragrance chemicals shown in Table V as well as many others showed insufficient activity toward either houseflies or mosquitoes to be considered effective semio chemicals.

TABLE V.

Nature-Identical Flavor/Fragrance Chemicals with Insignificant
Semio Activity Toward Houseflies and Mosquitoes

Linalool	Benzyl Acetate
Citral	Cyclopentadecanone
Nerol	Muscone
Geraniol	Ambrettolide
beta Phenethyl Alcohol	*delta* Decalactone
	2-Undecanone
Benzyl Alcohol	

On the other hand, several low volume usage flavor and fragrance chemicals including alcohols, ketones, esters, acids, and others have shown sufficient attractant or repellant activity to be classified as semio-active.

Some nature-identical materials with attractant activity are shown in Table VI. From this table it can be seen that only a few materials such as *alpha* terpineol, *beta* damascenone, *d*-pulegone, and *d*-carvone are relatively high usage flavor and fragrance items. The chemicals shown in Table VII, although not nature-identical, also show semio attractancy. It should be noted that, although most of these chemicals are odorous, none, at present, is used by the fragrance industry.

To this point, we have discussed chemicals which have shown semio attractancy. Now we shall describe several materials, both nature-identical and non-nature-identical, which function as repellants. Table VIII shows the nature-identical chemicals from this study which were found to be repellant. It should be noted that both *alpha* and *beta* damascones are repellants toward houseflies and mosquitoes whereas *beta* damascenone is an attractant for houseflies and moths. Even though all three chemicals possess a pronounced rose-apple odor, it is very interesting to observe how insects are very selective toward them.

TABLE VI.

Nature-identical Materials With Attractant Activity

Chemical (or Extract)	Natural Occurrence	Semio Attractant Toward
Alcohols:		
alpha Terpineol	Citrus Oils	Sandflies (*20*)
3-Methyl-3-buten-1-ol	Raspberry, Ylang	Houseflies (*21*)
n-Dodecanol	Eucalyptus Oil	Houseflies (*22*)
Ketones:		
d-Carvone	Caraway Oil	Houseflies (*23*)
		Mosquitoes (*24*)
		Beetles (*25*)
d-Pulegone	Pennyroyal Oil	Houseflies (*23*)
		Beetles (*25*)
beta Damascenone	Rose, Apple	Houseflies (*22*)
		Moths (*26*)
Esters and Acids:		
Ethyl 2-Methyl-3-pentenoate	Strawberry	Houseflies (*27*)
		Mosquitoes (*28*)
Benzyl Formate	Apple Blossom	Houseflies (*23*)
		Beetles (*25*)
*iso*Butyric Acid	Strawberry, Hops	Houseflies (*29*)
Miscellaneous:		
Methyl *iso*Eugenol	Orange, Nutmeg	Houseflies (*22*)
		Moths (*26*)
Dimethyl Disulfide	Pineapple, Cocoa	Sandflies (*30*)
		Houseflies (*30*)
		Mosquitoes (*31-32*)
Marigold Absolute	Marigold	Houseflies (*29*)

TABLE VII.

Non-nature-identical Chemicals with Attractant Activity

Chemical	Semio Attractant Toward
Alcohols:	
3-Ethyl-3-hexanol	Houseflies (*21*)
3-Ethyl-2-methyl-3-pentanol	Houseflies (*21*)
2,3-Dimethyl-3-hexanol	Houseflies (*21*)
9-Decen-1-ol	Houseflies (*21*)
10-Undecen-1-ol	Houseflies (*21*)
Esters:	
Dibutyl Succinate	Sandflies (*30, 33*)
	Mosquitoes (*31-32*)
	Beetles (*33*)
*iso*Amyl Decanoate	Houseflies (*29*)
sec. Undecyl Acetate	Houseflies (patent pending - IFF)
	Mosquitoes (patent pending - IFF)

TABLE VIII.

Nature-identical Chemicals with Repellant Activity

Chemical	Natural Occurrence	Semio Repellant Toward
Methyl Jasmonate	Jasmin, Boronia	Houseflies, Mosquitoes (patent pending - IFF)
Dihydro Methyl Jasmonate (*trans*) (Hedione)	Tea	Houseflies, Mosquitoes (patent pending - IFF)
epi Dihydro Methyl Jasmonate (*cis*)	Tea	Houseflies, Mosquitoes (patent pending - IFF)
alpha Damascone (racemic)	Tea	Houseflies, Mosquitoes (patent pending - IFF)
beta Damascone	Rose	Houseflies, Mosquitoes (patent pending - IFF)

Table IX shows the non-nature-identical chemicals which also show repellancy.

TABLE IX.

Non-nature-identical Chemicals with Repellant Activity

Chemical	Semio Repellant Toward
Alcohols and Ethers:	
1-Nonen-3-ol	Houseflies (*19, 34*)
1-Octen-4-ol	Houseflies, Mosquitoes (patent pending - IFF)
3,3,5,6,6-Pentamethyl-2-heptanol (Dihydro Koavol)	Houseflies, Mosquitoes (patent pending - IFF)
1-Ethyl-5-*iso*propoxy Tricyclo [2.2.1.0(2,6)]heptane (Isoproxen)	Houseflies, Mosquitoes (patent pending - IFF)
Ketones:	
3,4.5,6,6-Pentamethyl-3-hepten-2-one (Koavone)	Houseflies, Mosquitoes (patent pending - IFF)
trans, trans delta Damascone	Houseflies, Mosquitoes (patent pending - IFF)

We would now like to discuss some special fragrance chemicals which are known as Schiff bases. These are condensation products of aldehydes with methyl anthranilate. This type of compound is used extensively in the fragrance industry to provide a long-lasting effect to the fragrance. Table X shows some Schiff bases with repellant activity

Of these three, neither ethyl vanillin nor Lyral occurs in nature, but methyl anthranilate is an important component of fruits and flowers such as grape and citrus flowers.

Conclusion

In conclusion, we would like to say that highly odorous large volume flavor and fragrance chemicals such as linalool, geraniol, nerol, phenyl ethyl alcohol, benzyl acetate, cyclic musks and lactones, with the exception of citronellol, are of no semio value with respect to houseflies and mosquitoes. However, we have found that some molecules which are known to be key aroma character-donating components of naturals such as *d*-carvone in caraway, *d*-pulegone in pennyroyal, *beta* damascenone in rose, and methyl *iso*eugenol in

TABLE X.

Schiff Bases with Repellant Activity

Schiff Base	Repellant Toward
Vanillin-Methyl Anthranilate	Houseflies, Mosquitoes (patent pending - IFF)
Ethyl Vanillin-Methyl Anthranilate	Houseflies, Mosquitoes (patent pending - IFF)
4-(4-Hydroxy-4-methylpentyl)-3-Cyclohexen-1-carboxaldehyde (Lyral)-Methyl Anthranilate	Houseflies, Mosquitoes (patent pending - IFF)

nutmeg are attractants for houseflies, whereas *alpha* damascone from tea, *beta* damascone from rose and jasmonates from jasmin, tea, and boronia act as repellants. At the same time, very sweet-floral synthetic molecules such as Schiff bases are also found to be repellant.

From our studies, we conclude that there is no obvious odor-structure relationship between flavor and fragrance molecules and their semio activity, at least in the case of insects such as houseflies and mosquitoes.

Literature Cited

(1)   Jacobson, M.; Keiser, I.; Chambers, D.L.; Miyashita, D.H.; Harding, C. *J. Med. Chem.* **1971**, *14*, 236.
(2)   Jacobson, M.; Keiser, I.; Miyashita, D. H.; Harris, E. J. *Lloydia* **1976**, *39*, 412.
(3)   Hwang, Y.S.; Mulla, M.S.; Axelrod, H. J. *Chem. Ecol.* **1978**, *4*, 463.
(4)   Pierce, H.D., Jr.; Vernon, R.S.; Borden, J. H.;Oehlschlager, A.C. *J. Chem. Ecol.* **1978**, *4*, 65.
(5)   Cantelo, W.W.; Jacobson, M. *Environ. Entomol.* **1979**, *8*, 444.
(6)   Seifert, R.M. *J. Ag. Food Chem.* **1981**, *29*, 647.
(7)   Jacobson, M. *Econ. Bot.* **1982**, *36*, 346.
(8)   Jain, S.C.; Nowicki, S.; Eisner, T.; Meinwald, J. *Tet. Ltrs.* **1982**, *23*, 4639.
(9)   Inazuka, S. *Nippon Noyaku Gakkaishi* **1982**, *7*, 145.
(10)   Scriven, R.; Meloan, C. E. *Ohio J. Sci.* **1984**, *84*, 82.
(11)   Verma, M.; Meloan, C.E. *Amer. Lab. (Fairfield, Conn.)* **1981**, *13*, 64.
(12)   Richards, A.G., Jr.; Cutkomp, L.K. *J. N.Y. Entomol. Soc.* **1945**, *53*, 313.
(13)   *C&E News* Dec. 6, **1982**, 7.
(14)   Baker, T. C.; Nishida, R.; Roelofs, W. L. *Science* **1981**, *214*, 1359.

(15) Snow, J. W.; Coppedge, J.R.; Broce, A.B.; Goodenough, J.L.; Brown, H.E. *Bull. of the ESA*, **1982**, *28*, 277.

(16) Plimmer, J.R.; Inscoe, M.N.; McGovern, T.P. *Ann. Rev. Pharmacol. Toxicol.* **1982**, *22*, 297.

(17) Lee, S.; Chen, Y. *Nippon Noyaku Gakkaishi* **1977**, *2*, 135.

(18) Vale, G.A. *Bull. Entomol. Res.* **1980**, *70*, 563.

(19) Butler, J.F.; Katz, I. U.S. Patent 4,759,228 dtd July 26, 1988, assigned to International Flavors & Fragrances, New York, N.Y. and The University of Florida, Gainesville, Fla.

(20) Wilson, R.A.; Butler, J.F.; Withycombe, D.A.; Mookherjee, B.D.; Katz, I.; Schrankel, K.R. U.S. Patent 4,886,662 dtd December 12, 1989, assigned to International Flavors & Fragrances, New York, N.Y. and The University of Florida, Gainesville, Fla.

(21) Wilson, R.A.; Mookherjee, B.D.; Butler, J.F.; Withycombe, D.A.; Katz, I.; Schrankel, K.R. U.S. Patent 4,764,367 dtd August 16, 1988, assigned to International Flavors & Fragrances, New York, N.Y. and The University of Florida, Gainesville, Fla.

(22) Wilson, R.A.; Butler, J.F.; Withycombe, D.A.; Mookherjee, B.D.; Katz, I.; Schrankel, K.R. U.S. Patent 4,801,446 dtd January 31, 1989, assigned to International Flavors & Fragrances, New York, N.Y. and The University of Florida, Gainesville, Fla.

(23) Wilson, R.A.; Butler, J.F.; Withycombe, D.A.; Mookherjee, B.D.; Katz I., Schrankel, K.R. U.S. Patent 4,988,508 dtd January 29, 1991, assigned to International Flavors & Fragrances, New York, N.Y. and the University of Florida, Gainesville, Fla.

(24) Wilson, R.A.; Butler, J.F.; Withycombe, D.A.; Mookherjee, B.D.; Katz, I.; Schrankel, K.R. U.S. Patent 4,970,068 dtd November 13, 1990, assigned to International Flavors & Fragrances, New York, N.Y. and The University of Florida, Gainesville, Fla.

(25) Wilson, R.A.; Butler, J.F.; Withycombe, D.A.; Mookherjee, B.D.; Katz, I.; Schrankel, K.R. U.S. Patent 4,992,270 dtd February 12, 1991, assigned to International Flavors & Fragrances, New York, N.Y. and The University of Florida, Gainesville, Fla.

(26) Wilson, R.A.; Butler, J.F.; Withycombe, D.A.; Mookherjee, B.D.; Katz, I.; Schrankel, K.R. U.S. Patent 4,859,463 dtd August 22, 1989, assigned to International Flavors & Fragrances, New York, N.Y. and The University of Florida, Gainesville, Fla.

(27) Wilson, R.A.; Butler, J.F.; Withycombe, D.A.; Mookherjee, B.D.; Katz, I.; Schrankel, K.R. U.S. Patent 4,808,403 dtd February 28, 1989, assigned to International Flavors & Fragrances, New York, N.Y. and The University of Florida, Gainesville, Fla.

(28) Wilson, R.A.; Butler, J.F.; Withycombe, D.A.; Mookherjee, B.D.; Katz, I.; Schrankel, K.R. U.S. Patent 4,816,248 dtd March 28, 1989, assigned to International Flavors & Fragrances, New York, N.Y. and The University of Florida, Gainesville, Fla.

(29)   Wilson, R.A.; Butler, J.F.; Withycombe, D.A.; Mookherjee, B.D.; Katz, I.; Schrankel, K.R. U.S. Patent 4,988,507 dtd January 29, 1991, assigned to International Flavors & Fragrances, New York, N.Y. and the University of Florida, Gainesville, Fla.

(30)   Wilson, R.A.; Butler, J.F.; Withycombe, D.A.; Mookherjee, B.D.; Katz, I.; Schrankel, K.R. U.S. Patent 4, 801,448 dtd January 31, 1989, assigned to International Flavors & Fragrances, New York, N.Y. and The University of Florida, Gainesville, Fla.

(31)   Wilson, R.A.; Butler, J.F.; Withycombe, D.A.; Mookherjee, B.D.; Katz, I.; Schrankel, K.R. U.S. Patent 4,818,525 dtd April 4, 1989, assigned to International Flavors & Fragrances, New York, N.Y. and The University of Florida, Gainesville, Fla.

(32)   Wilson, R.A.; Butler, J.F.; Withycombe, D.A.; Mookherjee, B.D.; Katz, I.; Schrankel, K.R. U.S. Patent 4,902,504 dtd February 20, 1990, assigned to International Flavors & Fragrances, New York, N.Y. and The University of Florida, Gainesville, Fla.

(33)   Wilson, R.A.; Butler, J.F.; Withycombe, D.A.; Mookherjee, B.D.; Katz, I.; Schrankel, K.R. U.S. Patent 4, 911,906 dtd March 27, 1990, assigned to International Flavors & Fragrances, New York, N.Y. and The University of Florida, Gainesville, Fla.

(34)   Wilson, R.A.; Butler, J.F.; Withycombe, D.A.; Mookherjee, B.D.; Katz, I.; Schrankel, K.R. U.S. Patent 4, 696,676 dtd September 29, 1987, assigned to International Flavors & Fragrances, New York, N.Y. and The University of Florida, Gainesville, Fla.

RECEIVED August 19, 1992

Chapter 5

Formation of Some Volatile Components of Tea

Akio Kobayashi, Kikue Kubota, and Motoko Yano

Laboratory of Food Chemistry, Ochanomizu University, 2–1–1, Ohtsuka, Bunkyo-ku, Tokyo, Japan

A cluster analysis showing the complex gas chromatograms of tea volatiles correlates to aroma character, which is the main factor in classifying various teas. Some of the main volatiles were formed by enzymatic hydrolysis of the nonvolatile fraction from a hot-water extract of green tea, from which two glycosides were separated and identified as ß-D-glucosides. Optical isomers of linalool and 3,7-dimethyl-1,5,7-octatrien-3-ol were present with different R-,S-ratios among the various tea volatiles. This suggests different formation mechanisms among these structurally related compounds.

Tea is probably the most popular beverage all over the world. When we use the word "tea", it means a product from only one plant species, *Camellia sinensis*. With development of the tea drinking custom, various types of tea have been manufactured in different countries. According to their manufacturing processes, tea can be classified into three types, i.e., fermented, semifermented and non-fermented, which are generally called black, oolong and green teas, respectively. The fermentation process in tea manufacturing does not mean microbial fermentation, but involves changes in the taste, aroma and color by an enzymatic action in tea leaves. For the production of Japanese green tea, freshly plucked tea leaves are steamed to stop enzyme activity; therefore, its aroma resembles the fresh green plant. By crushing and tearing withered tea leaves, the various enzyme activities cooperate to produce the color, taste and aroma of black tea. Contrary to these two extreme processes, oolong tea is manufactured under milder fermentation conditions which proceed in keeping tea leaves almost intact through the withering and rolling processes. Therefore, "semi-fermented" does not mean an incomplete fermentation process (1) in black tea, but involves a different enzymatic action to produce the characteristic oolong tea aroma.

0097–6156/93/0525–0049$06.00/0

Table I. Tea Sample Used in a Clustar Analysis

Sample Number		Country	Commercial Name or Cultiver	Sample
Black Tea	B 1 - B 8	Sri Lanka	Ceilon Tea	8
	B 9 - B12	India	Darjeeling Tea	4
	B13	Japan	Benihikari	1
Oolong Tea	O14 - O23	China	Oolong Tea	10
	O24 - O27	Taiwan	Oolong Tea	4
	O28 - O29	Taiwan	Pouchong Tea	
Green Tea	G30 - G31	Japan	Sencha	2
	G32	Japan	Gyokuro	1
	G33 - G38	Japan	Sencha	6
	G39 - G44	Japan	Aracha*	5

                                                        Total   44

* made from five different cultivars

Cluster Analysis of the Tea Volatiles

The development of analytical techniques has made possible the identification of more than 300 compounds as tea volatile constituents (2), and studies on these volatiles have been directed to correlate such complicated constituents to their aroma characters. The chemometric approach has often been applied to correlate chemical analysis data to a sensory evaluation of some food products. Tea is thought to be a good example for the application of this method because there are some characteristic commercially available products which have been internationally evaluated. As shown in Table I, 44 samples of typical black, oolong and green teas were collected from their producing countries for a multivariate analysis, the tea aroma volatiles being isolated under the same conditions by simultaneous distillation and extraction apparatus, and then analysed by high-resolution gas chromatography (GC) and gas chromatography-mass spectrometry (GC-MS).

Comparison of the Gas Chromatograms. The gas chromatograms of the volatiles from three different teas are shown in Fig 1. Green tea shows the simplest gas chromatogram with about 150 separate peaks. The most complicated one is that of oolong tea with more than 250 peaks, and the gas chromatogram of black tea is characterized by the presence of several main peaks. However, almost all the peaks appearing on the green tea gas chromatogram are commonly present on the other two. Therefore, it is possible to apply a multivariate analysis to these gas chromatographic data.

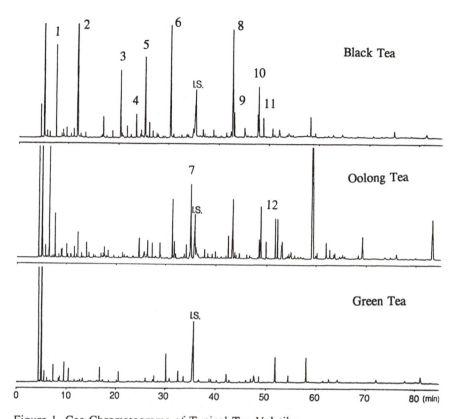

Figure 1. Gas Chromatograms of Typical Tea Volatiles
Gas chromatographic conditions: 50m x 0.25mm FS-WCOT column coated with PEG 20M. Programming of oven temperature; 60°C(4min hold) up to 180°C programmed with 2°C/min. Carrier gas; Nitrogen flow rate 1.2ml/min, Split ratio; 33:1. Compounds: 1.hexanal, 2.(Z)-2-hexenal, 3.(Z)-3-hexenol, 4.linalool oxide I, 5.linalool oxide II, 6.linalool, 7.3,7-dimethyl-1,5,7-octatrien-3-ol, 8.linalool oxide IV, 9.methyl salicylate, 10.geraniol, 11.benzyl alcohol, 12.phenylethanol.

Grouping of Commercial Tea by a Cluster Analysis. We picked up 77 peaks present commonly throughout the gas chromatograms, and the ratios of their peak areas to the internal standard (hexadecane) were used as variables for the analysis. The cluster analysis shown in Fig 2 (Togari,N., Ochanomizu University, Master Thesis) indicates that the GC pattern can clearly classify three different types of tea and, moreover, the grouping of a distinct tea can be correlated to a sensory evaluation or empirical knowledge about commercial tea. For example, black tea has two semi-clusters containing mainly Assam and Darjeeling teas, respectively. Green tea also has two major semi-clusters, and one of them is grouped very tightly. This implies that the green tea aroma is similar among commercially available products, and this similarity can be explained by the fact that it is impossible to change the aroma of non-fermented tea during the manufacturing process. Contrary to these two different types of tea, there are three semi-clusters

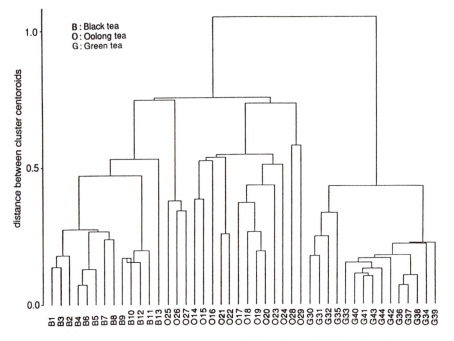

Figure 2. Dendrogram from Centroid Clustering of 44 Tea Samples

in oolong tea, two being Taiwan teas and one a product from the mainland of China; however, they are loosely connected to each other. This result shows that each oolong tea has a characteristic aroma through the various conditions of the manufacturing process.

By comparing the gas chromatograms of various tea volatiles, the yield of green tea is the lowest and the dynamic range of each peak area is small. Therefore, green tea can be expected to contain many precursors of tea volatiles, because the enzymatic action is stopped at an early manufacturing stage. The major peaks in black tea are presumed to be formed during the fermentation process by the enzymatic action in tea leaves from the corresponding precursor.

Formation of Tea Volatiles

If a non-volatile green tea extract is incubated with the crude enzymes of fresh tea leaves, some volatiles in the fermented tea should be freed from such precursors. A hot-water extract of green tea was distilled under reduced pressure and, from the distillate, the green tea volatiles were extracted and analysed by GC and GC-MS according to the usual methods. The residue from distillation was treated with the crude enzyme system which was prepared by crushing fresh tea leaves in liquid nitrogen and washing it with acetone (expressed as acetone powder hereafter) and the newly formed volatiles were extracted and analysed by the same steam distillation procedure described above.

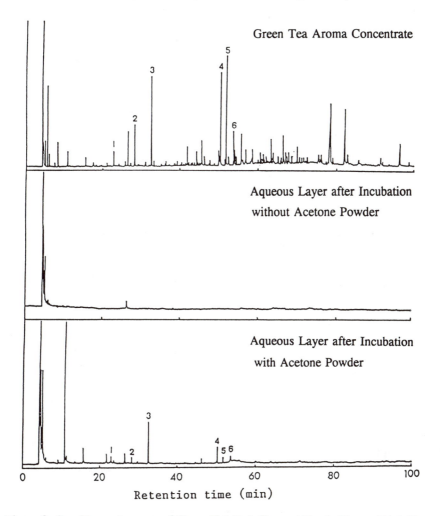

Figure 3. Gas Chromatograms of Green Tea Volatiles and Newly Formed Volatiles from Aqueous Layer.   Gas Chromatographic Conditions: Same as Figure 1. Compounds: 1.(Z)-3-hexenol, 2.linalool oxide I, 3.linalool, 4.geraniol, 5.benzyl alcohol, 6.phenylethanol.

Hydrolysis of the Non-volatile Precursors (*3*).   There are three gas chromatograms in Fig. 3 comparing green tea volatiles and the extracts from the aqueous solution after incubating with and without the acetone powder treatment. The newly formed compounds after treating with acetone powder were (Z)-3-hexenol, linalool oxide I (cis-furanoid type), linalool, geraniol, benzyl alcohol and phenylethyl alcohol, which also appear as the main peaks in the upper gas chromatogram. An increase in the amount of these compounds is thought to be one of the factors that characterizes the fermented tea aroma.

    As volatile alcohols are known to be present as glycosides in many kinds of fruit and other parts of plants (*4*), we hydrolyzed the water extract after steam distillation with commercially available ß-glucosidase or hydrochloric acid.   As

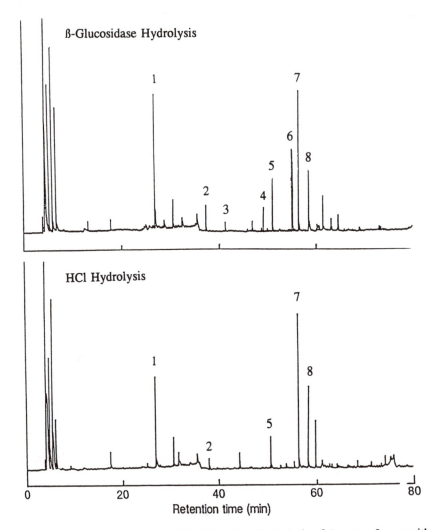

Figure 4. Gas Chromatograms of Volatiles after Hydrolysis of Aqueous Layer with Enzymes and Mineral Acid. Gas Chromatographic Condition: Same as Figure 1 except carrier gas flow rate; 1.0ml/min. Compounds: 1.(Z)-3-hexenol, 2.linalool, 3.3,7-dimethyl-1,5,7-octatrien-3-ol, 4.linalool oxide III, 5.methyl salicylate, 6.geraniol, 7.benzyl alcohol, 8.phenylethanol.

shown in Fig. 4, the gas chromatogram after treating with ß-glucosidase is more complicated than that obtained with the acetone powder, and the peaks of (Z)-3-hexenol and benzyl alcohol are significantly increased. Additional main peaks were identified as 3,7-dimethyl-1,5,7,-octatrien-3-ol, linalool oxide III (cis-pyranoid type) and methyl salicylate, all of which are also main components in semi- and fermented teas. Hydrolysis with hydrochloric acid gave the same results, but the gas chromatograms are simpler because the terpene alcohols were converted to other compounds under strongly acidic conditions.

Separation and Identification of the Glucosides. The increase in (Z)-3-hexenol and benzyl alcohol by hydrolysis with ß-glucosidase implied that both compounds were present as glucosides in the green tea extract. We isolated the glycosidic fraction from a hot-water extract of green tea with an XAD-2 column, and the glycosidic fraction thus obtained was then thoroughly acetylated to separate each glycoside by silica gel column chromatography. Two acetyl glycosides were separated in a pure state, which were identified as tetraacetates of (Z)-3-hexenyl- and benzyl-ß-D-glucopyranoside. The latter was a crystalline product with mp 99°C, and the spectrometric data were identical with those of an authentic sample. The chemical structure of the former was deduced from its NMR spectrum, and particularly, the (Z)-3- configuration of the aglycone was established by comparing to the split patterns of standard (Z)-3-hexenol from PMR. In conclusion, benzyl alcohol and (Z)-3-hexenol combined with glucose, and these glucosides have an important role as precursors during tea volatiles formation because they are known to increase during the fermentation process. In our experiment just described, the ß-glucosidase treatment of an aqueous extract of green tea increased the production of these two compounds. It is noteworthy that (Z)-3-hexenol has a greenish odor with a relatively low threshold value, and this is thought to be the origin of the green leaf odor because of its wide distribution in green plants. The biosynthetic route of (Z)-3-hexenol from linolenic acid with lipoxygenase and lyase has been established by Hatanaka's group in Japan (5); however, such a degraded product seems to be detoxicated and preserved in the form of a glycoside in plant tissue. The practical formation of (Z)-3-hexenol and its derivatives during tea processing should be the result of enzymatic degradation and hydrolysis of the glycosides.

Chirality of the Tea Volatiles

Among the tea volatiles, linalool is the most important, because the amount of linalool and its oxidative derivatives, linalool oxides, is 20% or more of the total volatiles of fermented teas (5). The structurally related compound, 3,7-dimethyl-1,5,7-octatrien-3-ol (abbreviated to trienol hereafter) was also identified after the glucosidase treatment of a glycosidic extract from green tea. Both compounds have an asymmetric carbon, and R- and S-linalools show different aroma characters. Therefore, the chirality of these compounds is important not only for their biochemical interconversion in plants or during tea fermentation, but also for clarifying the natural plant aroma character.

Preparation and Resolution of the Chiral Isomers. 3-R-linalool was obtained from natural Ho-oil, from which the 3R-trienol was derived by following the known method (7). Similarly, the racemic trienol was derived from racemic linalool, and these were used as standards for a gas chromatographic resolution of the chiral compounds in a CP column coated with permethylated ß-cyclodextrin as an optically active stationary phase. The gas chromatographic conditions are as follows: 25m x 0.25mm fused silica OT column, Detector; FID, Oven Temp.;60°C(for 8 min) up to 200°C with 1°C/min, Carrier gas; He, 1 ml/min. The R-isomer of both linalool and trienol was eluted earlier than the S-isomer, and their R,S-ratio calculated from the peak areas is summarized in Table II with the Kovats Indices. The R,S-ratio of linalool in different types of tea is almost the same. On the other hand, only S-trienol was identified in the oolong and black tea volatiles.

Table II. R,S-Ratio of Linalool and Trienol in Tea

		Black Tea	Oolong Tea	Green Tea	KI
Linalool	R	31.3	35.4	44.3	1252
	S	68.7	64.6	55.7	1254
Trienol	R	0	0	—	1273
	S	100	100	—	1278

Trienol in green tea was too little to identify on a gas chromatogram under the same conditions. The presence of S-trienol in black tea after using the chemical method (*8*) to synthesize the optical isomers and measuring the optical rotation has already been reported. In this experiment, the earlier conclusion was confirmed by the gas chromatographic resolution of a trace amount directly from the natural product. It is noteworthy that the difference of R,S-ratio between linalool and trienol suggests that the direct biogeneration of trienol from linalool is not plausible.

Conclusion

The attractiveness of volatile compounds is the most important factor in food flavor. However, the constituents of volatiles are highly complex, the volatiles of tea being no exception. To understand such complexity of analytical data and to correlate them to human taste, we have to adopt many different approaches that incorporate chemistry, biochemistry, physiology and mathematics.

As an example of the unification of different approaches, we have analyzed the attractiveness of tea volatiles by a multivariate analysis, identification of the precursors, and chiral compound resolution.

Literature Cited

1.   Kobayashi,A.; Tachiyama,K.; Kawakami,M.; Yamanishi,T.; Juan,I-M.; Chiu,W.T.F. *Agric.Biol.Chem.* **1984**, *49*, 1655-1660.
2.   Flament,I. *Food Review Int.* **1989**, *5*, 317-414.
3.   Yano,M.; Okada,K.; Kubota,K.; Kobayashi,A. *Agric.Biol.Chem.* **1990**, *54*, 1023-1028.
4.   *Bioflavour*; Schreier,P., Ed.; de Gruyter: Berlin, New York, **1988**, 584 pp.
5.   Hatanaka,A.; Kajiwara,T.; Sekiya,J.; Imoto,M.; Inoue,S. *Plant Cell Physiol.* **1982**, *23*, 91-99.
6.   Kobayashi,A.; Kawakami,M. In *Essential Oil and Waxes*; Linskens,H.F.; Jackson,J.F.O, Eds; Modern Methods of Plant Analysis;Springer-Verlag: Berlin, **1991**, Vol.12; 21-40.
7.   Felix,D.; Merela,A.; Seibl,J.; Kovats,E.sz. *Helv.Chim.Acta.* **1963**, *46*, 1513-1536.
8.   Nakatani,Y.; Sato,S.; Yamanishi,T. *Agric.Biol.Chem.* **1969**, *33*, 967-968.

RECEIVED August 19, 1992

Chapter 6

Antimicrobial Activity of Green Tea Flavor Components

Effectiveness against *Streptococcus mutans*

I. Kubo

Division of Entomology and Parasitology, College of Natural Resources, University of California, Berkeley, CA 94720

The antimicrobial activity of the 10 most abundant volatile components of green tea flavor (**1-10**) was examined. The activity of each volatile was moderate but broad in spectrum. Most of the volatiles tested inhibited the growth of one of the most important cariogenic bacteria, *Streptococcus mutans*. Among them, nerolidol (**4**) was the most potent; linalool (**1**) was the least effective. In addition, indole (**7**) significantly enhanced the activity of δ-cadinene (**2**) and β-caryophyllene (**10**) against *S. mutans*. These two sesquiterpene hydrocarbons also showed potent activity against a dermatomycotic bacterium, *Propionibacterium acnes*. Lastly, but most importantly, indole inhibited the growth of all of the Gram-negative bacteria tested, *Pseudomonas aeruginosa*, *Enterobacter aerogenes*, and *Escherichia coli*.

Tea is one of the most widely consumed beverages in the world. Its popularity is attributed to its pleasant flavor, combined with its stimulating effects. There are many types of tea, including green tea, black tea and oolong tea, each with several subclassifications (*1*). All are prepared from what is basically the same plant, *Camellia sinensis* L. (Theaceae) and varies by different manufacturing processes.

It has been said that those who continuously drink large amounts of green tea have less tooth decay. This old tradition was proven by Onishi *et al* after a year of continuous surveillance at elementary schools (*2*). This group also reported that green tea extract contained many active substances for cavity prevention (*3*). The active principles have not yet been thoroughly defined, although several polar polyphenolic compounds in the green tea have already been reported as moderate antibacterial principles (*4*) against *Streptococcus mutans*, which is the primary bacterium responsible for causing dental cavities in experimental animals and humans (*5*). The minimum inhibitory concentrations (MICs) of these polyphenols, reported against *S. mutans* were at most 250 µg/ml (*4*). Theoretically, dental cavities can be prevented by eliminating *S. mutans*. This cariogenic bacterium

0097–6156/93/0525–0057$06.00/0

adheres firmly to smooth tooth surfaces and produces sticky, water-insoluble glucans from dietary sucrose that facilitates the accumulation of the other oral microorganisms. *S. mutans* and these microorganisms form plaque on enamel, developing cavities (Figure 1) (5-7). Although the popularity of green tea is attributed to its pleasant flavor combined with its stimulating effects, the antimicrobial activity of non-polar substances, particularly volatile flavor compounds, have not yet been investigated. Therefore, the antibacterial activity of the 10 major flavor constituents of green tea against *S. mutans* was examined.

With the pressing need for new antimicrobial agents in cosmetics, the same flavor compounds were tested against 12 other microorganisms (8-10). The lack of effective preservatives to control microorganisms which putrefy nutritious cosmetic products is a major problem to be solved. Control of specific microorganisms which cause skin, hair and oral problems is becoming even more important. In contrast to medicines which are used to heal ill people, antimicrobial agents in cosmetics are repeatedly applied to healthy skin, hair and teeth, often for long periods, which makes the safety of these products the primary importance. It is possible that edible plants, daily beverages, and food spices may be a superior source of new antimicrobial agents (8,9).

Flavor Compounds Tested (Figure 4)

The complex green tea flavor contains over 100 volatile compounds (11,12). The 10 most abundant volatile flavor constituents identified in green tea, namely: linalool (**1**), δ-cadinene (**2**), geraniol (**3**), nerolidol (**4**), α-terpineol (**5**), *cis*-jasmone (**6**), indole (**7**), ß-ionone (**8**), 1-octanol (**9**) and β-caryophyllene (**10**), in decreasing concentration, were selected for the assay from the list reported previously (13). Most teas are of similar composition; their compositions varies as a result of different manufacturing processes (12-15). Furthermore, these same volatile compounds are identified in many edible plants, food spices and beverages, and are frequently used for fragrances and flavors (16). For example, the most abundant component in green tea flavor, linalool (**1**), was also found in food spices such as coriander, lavender, sage, thyme, *etc.* (17-19), often, as the main component. Recently, green tea flavor itself has been used in ice cream, candy, soft drinks, *etc.* There is no doubt that these volatile compounds have long and widely been consumed by many people.

Assay Method Employed

Before antimicrobial activity of an individual compound is discussed, it should be emphasized that the broth dilution method (20) was used throughout this experiment, since these non-polar flavor compounds tested are not soluble in water. As a matter of fact, they did not show any activity by the paper disk method, since these water insoluble compounds might not diffuse into the media and/or because these volatiles were partially or even entirely evaporated from the paper disk when the solvent was removed prior to the assay.

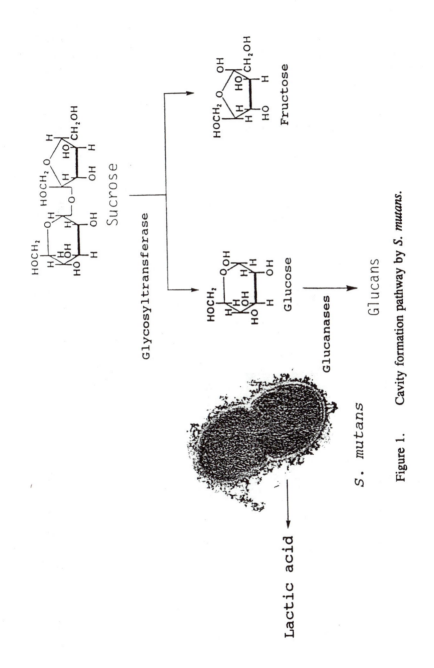

Figure 1. Cavity formation pathway by *S. mutans*.

Table I. Antimicrobial Activity of Green Tea Flavor Compounds

Microorganisms Tested	MIC (μg/ml)									
	1	2	3	4	5	6	7	8	9	10
Gram-positive bacteria										
Bacillus subtilis ATCC 9372	800	50	400	25	800	800	400	100	400	50
Brevibacterium ammoniagenes ATCC 6872	800	50	400	25	>800	>800	800	100	400	100
Staphylococcus aureus ATCC 12598	>800	>800	800	50	>800	>800	400	200	800	>800
Streptococcus mutans ATCC 25175	1600	800	400	25	400	800	800	100	400	>1600
Propionibacterium acnes ATCC 11827	200	3.13	400	25	100	400	200	25	200	6.25
Gram-negative bacteria										
Pseudomonas aeruginosa ATCC 10145	>800	>800	>800	>800	>800	>800	800	>800	>800	>800
Enterobacter aerogenes ATCC 13048	>800	>800	>800	>800	>800	>800	800	>800	>800	>800
Escherichia coli ATCC 9637	>800	>800	800	>800	800	>800	800	>800	400	>800
Yeasts										
Saccharomyces cerevisiae ATCC 7754	800	>800	400	>800	800	800	>800	>800	400	>800
Candida utilis ATCC 13048	400	>800	400	>800	800	800	>800	400	200	>800
Pityrosporum ovale ATCC 14521	400	>800	200	800	400	200	200	>800	100	>800
Molds										
Penicillium chrysogenum ATCC 10106	800	>800	200	800	400	200	50	400	200	>800
Trichophyton mentagrophytes ATCC 18748	200	>800	200	12.5	200	200	100	50	200	>800

Antibacterial Activity Against S. *mutans*

The antimicrobial activity of the 10 selected flavor compounds (**1-10**) against the 13 selected microorganisms is listed in Table I (*21*). The compounds tested exhibited activity against S. *mutans*, with the exception of β–caryophyllene (**10**). Among them, nerolidol (**4**) was the most potent, with an MIC of 25 μg/ml, while linalool (**1**) was the least effective, with an MIC of 1600 μg/ml. Although the potency of each compound against S. *mutans* is moderate to weak, the results indicate that, in addition to the use of flavor, the total volatile compounds seem to possess an additional function, namely "anti cavity activity", to some degree.

The total activity of a cup of green tea was reported to be enough to control S. *mutans* (*2*). The yield of the volatile flavor compounds obtained by steam-distillation from 16 kg of green tea was reported at about 5.6 g (*13*). If this is the case, theoretically a cup of green tea prepared with 2 g (the usual amount for a commercial tea bag) of the tea leaves in 100 ml of hot water contains a total of 7 μg/ml of volatiles. This concentration does not seem to be strong enough to control S. *mutans*, even if the volatile component of tea was assumed to consist only of nerolidol (**4**), the most potent antibacterial substance against S. *mutans*, among the 10 flavor compounds tested.

Minimum Bactericidal Activity

The MIC of each flavor compound alone against this cariogenic bacterium may not fully explain Onishi's observation (*2,3*). Hence, the minimum bactericidal concentration (MBCs) of the 10 flavor compounds were obtained as previouly described by Pearson *et al.* (*22*). The result is listed in Table II. Their MBC to MIC ratios were no greater than two with the exception of nerolidol. Thus, the MBC was the same as the corresponding MIC in linalool, δ-cadinene, geraniol and 1-octanol. The MBC was two-fold higher than the coresponding MIC in α-terpineol, *cis*-jasmone, indole and β-ionone, and that of nerolidol was eight-fold higher.

Table II. Bactericidal Activity of 10 Major Green Tea Flavor Compounds against S. *mutans*

Compounds Tested	MBC (μg/ml)
Linalool (**1**)	1600
δ-Cadinene (**2**)	800
Geraniol (**3**)	400
Nerolidol (**4**)	200
α-Terpineol (**5**)	1600
Cis-Jasmone (**6**)	1600
Indole (**7**)	1600
β-Ionone (**8**)	200
1-Octanol (**9**)	400
β-Caryophyllene (**10**)	>1600

Combination Effects

The MIC also differs from the *in vivo* assay, especially from continuous drinking of a large amount of green tea. In addition, tea contains many other chemicals such as the aforementioned antibacterial polyphenolic compounds. The combination of these substances may synergize the total antibacterial activity against *S. mutans*.

Based on these concerns, an attempt to enhance the antibacterial activity against *S. mutans* was made through combination of two or more green tea flavor compounds. In the preliminary combination studies, indole (**7**), the most abundant nitrogen containing compound in green tea flavor, was found to enhance the antibacterial activity of several other green tea flavor compounds against *S. mutans*. Therefore, a detailed study with the 4 most abundant green tea flavor compounds (**1**-**4**), in combination with indole, was carried out by the broth checkerboard method (*23*). Table III shows the MICs of these compounds against *S. mutans* in combination with 400 µg/ml of indole (equivalent to ½MIC for this cariogenic bacterium). In this combination, the activity of three terpene alcohols, linalool (**1**), geraniol (**3**) and nerolidol (**4**) against *S. mutans* was not significantly increased. Their MICs were enhanced by only 2-4 fold. In contrast, the activity of δ-cadinene (**2**), the second most abundant compound in the green tea flavor, was synergized 126-fold. The MIC of this sesquiterpene hydrocarbon was lowered from 800 to 6.25 µg/ml. Based on this finding, the other sesquiterpene hydrocarbon in the green tea flavor, β-caryophyllene (**10**), was also tested in combination with indole. Although β-caryophyllene did not exhibit any activity against *S. mutans* up to 1600 µg/ml when it was tested alone, the activity of β-caryophyllene was enhanced more than 256-fold when tested in combination with indole; the MIC was lowered to 6.25 µg/ml. The enhancing activity of indole seems to depend on the chemical combinations of the individual compounds.

Table III. MICs (µg/ml) of Green Tea Flavor compounds alone and in combination with ½MIC of Indole (**7**) against *S. mutans*

Linalool (**1**)	1600 → 800
δ-Cadinene (**2**)	800 → 6.25
Geraniol (**3**)	400 → 200
Nerolidol (**4**)	50 → 12.5
β-Caryophyllene (**10**)	>1600 → 6.25

Growth Studies

The MIC values alone do not fully characterize the antibacterial activity of these volatile compounds. Therefore, a more detailed study of the two most abundant compounds in green tea flavor, linalool (**1**) and δ-cadinene (**2**), was carried out.

First, the growth of *S. mutans* was investigated by measuring both culture turbidity and enumerating viable cells. The growth curves of *S. mutans* in the presence of linalool alone and in combination with indole by measuring turbidity are illustrated in Figure 2 and 3. Linalool alone suppressed the growth of this bacterium at 1600 µg/ml, while 800 µg/ml showed little restriction of the growth. However, 800 µg/ml of linalool in combination with 400 µg/ml (equivalent to ½MIC against *S. mutans*) of indole suppressed the growth over a 48-hours period. Moreover, 400 µg/ml of linalool combined with 400 µg/ml of indole still increased the culture lag time to 24 hours, when the lag time for the culture containing linalool alone at 400 µg/ml was approximately 6 hours.

By measuring culture turbidity for the first 48 hours at 6-hour intervals, it was found that the concentrations of δ-cadinene, lower than 800 µg/ml, did not completely inhibit the growth but rather increased the lag time of *S. mutans* as shown in Figure 2 and 3. The lag time for a culture containing 1.56 µg/ml of δ-cadinene was 12 hours, with only a 6-hour delay from the lag time of the control culture. However, at a concentration of 3.13 µg/ml, δ-cadinene had a significant effect on *S. mutans* producing a lag time of 30 hours. Also, δ-cadinene at 6.25 and 12.5 µg/ml produced a lag time of 36 hours and concentrations of 25, 100 and 400 µg/ml restricted the final culture turbidity. Beside the inhibitory effect of δ-cadinene alone, its combination effect with indole was also noticeable. At 1.56, 3.13 and 6.25 µg/ml, δ-cadinene combined with 400 µg/ml of indole retarded the growth longer than the single treatment. From 12.5 to 400 µg/ml of δ-cadinene in combination, the growth was completely inhibited up to 72 hours.

Second, to confirm the finding of the combination study by measuring culture turbidity, time-kill curves were established. The killing effects of indole (**7**) and linalool (**1**), both alone and in combination, are illustrated in Figure 2 and 3. *S. mutans* tolerated up to 400 µg/ml of indole with little restriction of growth, while the cell numbers declined slowly in the presence of 800 µg/ml of indole. Linalool at a concentration of 1600 µg/ml proved bactericidal. 800 µg/ml of linalool suppressed the growth over 12 hours of incubation and very little reduction in growth was noted when 400 and 200 µg/ml of this compound were used. However, 800 µg/ml of linalool plus 400 µg/ml of indole (equivalent to ½MIC) showed bactericidal activity and 400 µg/ml of indole suppressed the growth, cell numbers increasing slowly over 32 hours of incubation.

In contrast, *S. mutans* was more sensitive to the combination of δ-cadinene and indole as shown in Figure 2 and 3. A lethal effect was seen when 800 µg/ml of δ-cadinene was employed. This compound, at a concentration of 400 µg/ml, produced an apparent decrease in the viable cell count of *S. mutans* over 32 hours of incubation. This was followed by an increase in cell numbers, to a final cell count that was at the same level as the control culture. Same type of growth pattern was obtained with 6.25 to 25 µg/ml of δ-cadinene. Thus, cell numbers of *S. mutans* declined slowly over 24 hours incubation and then rapid growth occurred. These recoveries of the cell growth were not observed in combination with 400 µg/ml of indole. Even 25 µg/ml of δ-cadinene combined with 400 µg/ml of indole proved bactericidal against *S. mutans*, while at the concentration of 6.25 µg/ml, final cell count was three log cycles lower than the initial inoculum.

Although we have previously described the potentiation of antifungal activity

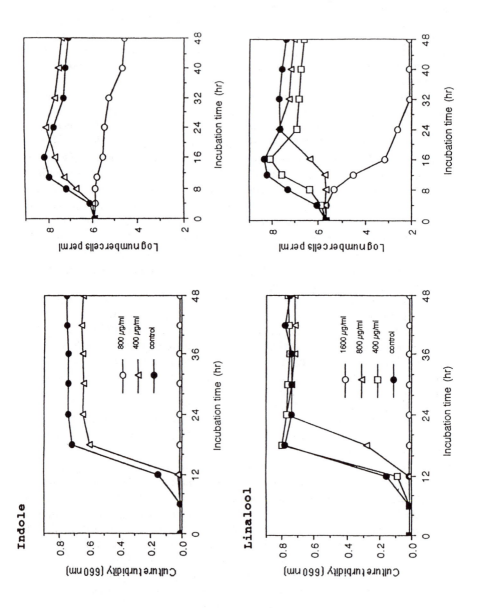

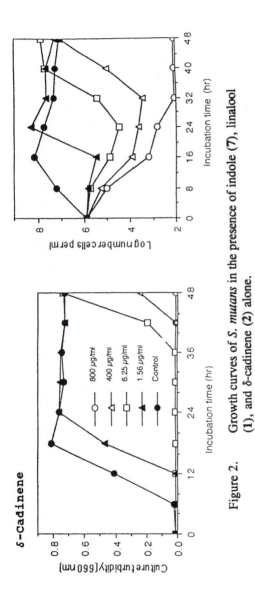

Figure 2. Growth curves of *S. mutans* in the presence of indole (**7**), linalool (**1**), and δ-cadinene (**2**) alone.

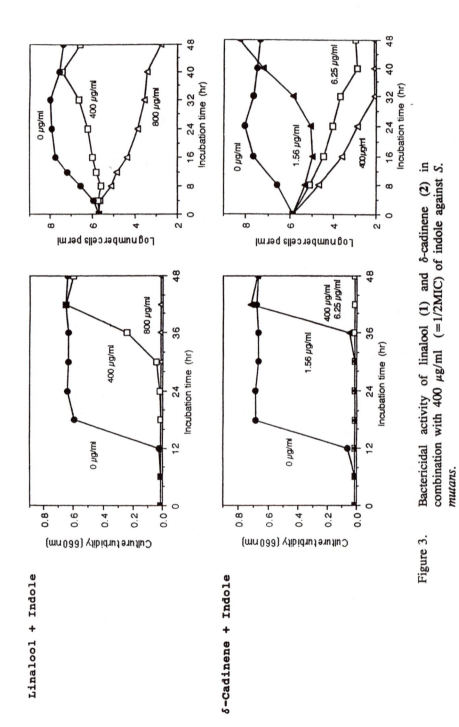

Figure 3.   Bactericidal activity of linalool (1) and δ-cadinene (2) in combination with 400 μg/ml (=1/2MIC) of indole against *S. mutans.*

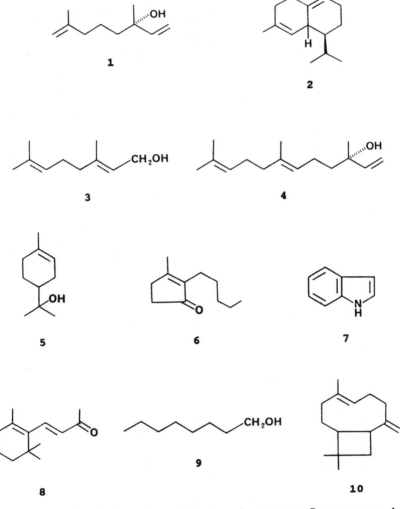

Figure 4. Chemical structures of the ten major green tea flavor compounds.

of several antibiotics, especially against *S. cerevisiae* and *C. utilis* (*24*), and *Candida albicans, Pityrosporum ovale* and *S. cerevisiae* (*8*), this is the first report of potentiation of antibacterial activity against *S. mutans* by combining two substances. Interestingly, a large quantity of indole is contained in jasmine which is sometimes added to tea for flavor. Jasmine may have this function, in addition to enriching the flavor.

Antimicrobial Activity Against Other Microorganisms

Besides the activity against *S. mutans,* activity against all other microorganisms tested was also shown by the 10 volatile compounds (**1-10**). Each compound exhibited moderate, but broad spectrum of antimicrobial activity. Among the microorganisms tested, *P. acnes* was the most sensitive, with MICs between 3.13-400 µg/ml. This bacterium is one of the bacteria responsible for acne, causing inflammation and comedos (*25*). The two most active compounds, δ-cadinene (**2**) and β-caryophyllene (**10**) may be particularly useful as protection from *P. acnes* infection. The activity of these sesquiterpene hydrocarbons against other microorganisms was mainly limited to Gram-positive bacteria, with MICs between 50-100 µg/ml. In addition, since the antimicrobial activity of δ-cadinene and β-caryophyllene against *S. mutans* was significantly synergized by indole, these two sesquiterpene hydrocarbons were also tested in combination with indole to examine if indole also had the same enhancing activity against *P. acnes*. The result, unexpectedly, was that indole did not exhibit any meaningful enhancing activity against this bacterium as shown in Table IV. The enhancing activity of indole seems to depend not only on the chemicals being combined, but also the test microorganisms.

Table IV. MICs of δ-Cadinene (**2**) and β-Caryophyllene (**10**) alone and in combination with ½MIC of Indole (**7**) against *P. acnes*

δ-Cadinene (**2**)	3.13 → 1.56
β-Caryophyllene (**10**)	6.25 → 3.13

In contrast to *P. acnes*, *S. aureus* was the least sensitive Gram-positive bacterium. Only nerolidol (**4**), ß-ionone (**8**), geraniol (**3**) and 1-octanol (**9**) showed some activity, with MICs of 250, 200, 800 and 800 µg/ml, respectively.

Most noticeable in this experiment, indole (**7**), exhibited antibacterial activity against all the Gram-negative bacteria tested, *P. aeruginosa, E. aerogenes* and *E. coli*. The MICs were 800, 800 and 400 µg/ml, respectively. The antibacterial activity of indole against several *Pseudomonas* species was previously reported; however, it was isolated from microbial fermentation (*26, 27*). Generally, few phytochemicals exhibit activity against Gram-negative bacteria, especially *Pseudomonas* species. Geraniol (**3**), α-terpineol (**5**) and 1-octanol (**9**) also showed weak activity against *E. coli*, with MICs of 800, 800 and 400 µg/ml, respectively.

In addition to antibacterial activity, most of the green tea flavor compounds tested exhibited antifungal activity against *P. ovale, S. cerevisiae, C. utilis, T. mentagrophytes* and *P. chrysogenum.* Most significantly, nerolidol (**4**) inhibited the growth of *T. mentagrophytes* at 12.5 µg/ml. All other volatiles except δ-cadinene and β-caryophyllene also inhibited the growth of *T. mentagrophytes,* with MICs between 50-200 µg/ml. This fungus occurs primarily on human hair causing human dermatomycosis. Similarly, the growth of another dermatomycotic fungus, *P. ovale,* was also inhibited by the same flavor components (*21*).

Conclusion

The green tea flavor compounds described may be considered as potential antimicrobial agents for cosmetic and food products (*18,19*). For example, linalool, β-caryophyllene and indole have long been used as food additives as listed in Table V.

Table V. Reported Uses of Linalool (**1**) and β-Caryophyllene*(**10**)

	Linalool (**1**)	β-Caryophyllene (**10**)
Non-alcoholic beverage	2.0	14**
Ice cream, ices, *etc.*	3.6	2.0
Candy	8.4	34
Baked goods	9.6	27
Chewing gum	0.9	200
Condiments	40	50

*, Fernaroli's Handbook of Flavor Ingredients; **, ppm

Antibiotics such as penicillin, erythromycin, tetracycline *etc.* effectively prevented dental cavities *in vitro* and *in vivo* (*28-30*), but they resulted in derangement of oral and intestinal bacterial floras. These are obviously undesirable and unacceptable side effects (*25*). The compounds identified in a common beverage such as tea, should not cause these undesirable side effects. Since green tea has been continuously consumed by many people for centuries, either the extract or purified flavor compounds of green tea might be considered safe for practical use such as in oral care products.

Acknowledgments

The work was conducted by Mr. H. Himejima and Ms. H. Muroi. Linalool, α-cadinene, nerolidol, α-terpineol, *cis*-jasmone, β-ionone, 1-octanol and β-caryophyllene were gifts from Takasago International Corporation.

Literature Cited

1. Eden, T. *Tea.* 3rd ed.; Longman, London, **1976**.
2. Onishi, M.; Shimura,; Nakamura, C.; Sato, M. *J. Dent. Hlth.* **1981**, *31*, 13-19.

3. Onishi, M.; Ozaki, F.; Yoshino, F.; Murakami, Y. *J. Dent. Hlth.* **1981**, *31*, 158-161.
4. Sakanaka, S.; Kim, M.; Taniguchi, M.; Yamamoto, T. *Agric. Biol. Chem.* **1989**, *53*, 2307-2311.
5. Hamada, S.; Slade, H. D. *Microbiol. Rev.* **1980**, *44*, 331-384.
6. de Jong, M. H.; Van der Hoeven, J. S.; Van Os, J. H.; Olijve,; J. H. *Appl. Environ. Microbiol.* **1984**, *47*, 901-904.
7. Walter, J. L. *Microbiol. Rev.* **1986**, *50* 353-380.
8. Kubo, I.; Himejima, M. *J. Agric. Food Chem.* **1991**, *39*, 2290-2292.
9. Kubo, I.; Himejima, M.; Muroi, H. *J. Agric. Food Chem.* **1991**, *39*, 1984-1986
10. Himejima, M.; Kubo, I. *J. Agric. Food Chem.* **1991**, *39*, 418-421.
11. Lament, I. Coffee, cocoa, and tea. in *Volatile Compounds in Foods and Beverages*, Ed. Maarse, H.; Dekker, New York, **1991**, pp. 617-669.
12. Yamaguchi, K.; Shibamoto, T. *J. Agric. Food Chem.* **1981**, *29*, 366-370.
13. Nose, M.; Nakatani, Y.; Yamanishi, T. *Agric. Biol. Chem.* **1971**, *35*, 261-271.
14. Kiribuchi, T.; Yamanishi, T. *Agric. Biol. Chem.* **1963**, *27*, 56-59.
15. Yamanishi, T.; Nose, M.; Nakatani, Y. *Agric. Biol. Chem.* **970**, *34*, 599-608.
16. Owuor, P.; Hirota, H.; Tsushida, T.; Murai, T. *Tea.* **1986**, *7*, 71-78.
17. Hazarika, M.; Mahanta, P. K.; Takeo, T. *J. Sci. Food Agric.* **1984**, *35*, 1201-1207.
18. Bauer, K.; Garbe, D.; Surburg, H. *Common Fragrance and Flavor Material, Preparation, Properties and Uses*; VCH Publishers, Weinheim, **1991**.
19. Maarse, H. *Volatile Compounds in Foods and Beverages*; Dekker, New York, **1991**.
20. Taniguchi, M.; Satomura, Y. *Agric. Biol. Chem.* **1972**, *36*, 2169-2175.
21. Kubo, I.; Muroi, H.; Himejima, M. *J. Agric. Food. Chem.* **1992**, *40*, 245-248.
22. Pearson, R. D.; Steigbigel, R. T.; Davis, H. T.; Chapman, S. W. *Antimicrob. Agents Chemother.* **1980**, *18*, 699-708.
23. Norden, C. W.; Wentzel, H.; Keleti, E. *J. Infect. Dis.* **1979**, *140*, 629-633.
24. Kubo, I.; Taniguchi, M. *J. Nat. Prod.* **1988**, *51*, 22-29.
25. Matsuoka, L. Y. *Pediatrics.* **1983**, *39*, 849-853.
26. Matsuda, K.; Toyoda, H.; Kakutani, K.; Hamada, M.; Ouchi, S. *Agric. Biol. Chem.* **1990**, *54*, 3039-3040.
27. Oimoni, M.; Hamada, M.; Hara, T. *J. Antibiot.* **1974**, *27*, 987-988.
28. Fitzgerald, R. J. *Antimicrob. Ag. Chemother.* **1972**, *1*, 296-302.
29. McClure, F. J.; Hewitt, W. L. *J. Dent. Res.* **1946**, *25*, 441-443.
30. Stephan, R. M.; Fitzgerald, R. J.; McClure, F. J.; Harris, M. R.; Jordan, H. V. *Dent. Res.* **1952**, *31*, 421-427.

RECEIVED October 19, 1992

ESSENTIAL OILS

Chapter 7

Essential Oils of the Eucalypts and Related Genera

Search for Chemical Trends

D. J. Boland[1] and J. J. Brophy[2]

[1]International Council for Research in Agroforestry, P.O. Box 30677, Nairobi, Kenya
[2]Department of Organic Chemistry, University of New South Wales, P.O. Box 1, Kensington, N.S.W. 2033, Australia

An examination of the essential oils of the genus *Eucalyptus* in the light of proposed splitting of the genus into a series of smaller genera is being undertaken. The purpose of the examination is to determine if there is any correspondence between the essential oils contained in the species and their place in the proposed new genera. To date, with 30% of the approximately 800 taxa examined, few trends in the essential oil data are apparent. Within *Eucalyptus* s. str. (encompassng some 170 taxa), Section Renantheria, a group of over 70 species containing *cis*- and *trans*-p-menth-2-en-1-ol and *cis*- and *trans*-piperitol, as well as piperitone has been idntified. These compounds are not found elsewhere in *Eucalyptus* s. lat. Although at present there has been a more thorough coverage of species in eastern Australia than in western Australia, it does appear that there is a much greater probability of finding aromatic compounds in species originating from Western Australia (and crossing proposed new generic boundaries), no doubt arising from the isolation of the western part of the continent in the Cretaceous period. Also crossing the proposed generic boundaries, it has been noticed that if oil yields are moderate to high (> 1%), 1,8-cineole is most likely to be the major component. If the oil yield is low then α-pinene is usually the principal component. In the course of this survey a significant number of species, whose essential oils differ markedly from the "typical" *Eucalyptus* oil, have been encountered. The composition of these oils is mentioned in the text.

Trees of the genus *Eucalyptus* (family Myrtaceae) are commonly called gum trees and comprise some 800 taxa (species/subspecies). The eucalypts are a dominant feature of the Australian vegetation with only two species not occurring naturally in Australia. All species have oil glands in their leaves, though not all of these glands contain oil. The percentage of oil contained in the fresh leaves varies

0097–6156/93/0525–0072$06.00/0

widely between species (0% to ~ 6% on fresh weight of leaf). Because of the large number of species, many eucalypt taxonomists have sought over the past 100 years to define taxonomic sub-grouping within the genus in order to better reflect their evolutionary histories. This clearly implies that some species are more closely related than others. Indeed, some scholars have felt that the genus is "unnatural" and really comprises several smaller genera deserving taxonomic recognition. These scholars also felt that several currently established genera clearly related to eucalypts should also be included in a larger scheme of species groupings to better reflect evolutionary relationships.

The most recent broad-scale, but still informal, classificatory commentary on the genus was by Johnson and Briggs in 1983 (*1*), based on a previous work by Pryor and Johnson (*2*). They proposed that a single tribe Eucalypteae should be erected for eucalypts and allied genera comprising four groups *(Arillastrum, Angophora, Eudesmia* and *Eucalyptus* groups) which in turn contain respectively 4, 3, 4 and 4 genera of unequal size. An evolutionary-implied cladogram of these divisions is shown in Figure 1. This scheme is based entirely on selected morphological characters (both vegetative and floral). It takes no account of essential oils. The cladogram (or tree) reflects, through its relative lengths of branches, the evolutionary closeness (or otherwise) of genera within the Tribe Eucalypteae. By no means should this cladogram be considered the final word in the eucalypt classification debate but it does provide a sensible framework for a relative assessment of the essential oils in each of the major groupings. It is proposed to discuss the essential oils of species within this framework from information currently available. So far only about 30% of the species have been rigorously screened for oils but some information of species from all groups is available.

The aim of this paper is, therefore, to summarize the main essential oil results to date within the Johnson & Briggs (1983) taxonomic framework (*1*) and attempt to search for trends in the chemistry of the essential oils within this framework. This paper also indicates species groups for which oil chemistry information is still required.

What follows is a summary of the main essential oil components found based on subtribe, genus, section and series following the sequence indicated in Figure 1. Many of these section and series names have never been formally recognised through publication in appropriate journals and are used here really as reference groupings of species. These groupings are understood by practicing taxonomists familiar with the family.

Arillastrum Group

The *Arillastrum* group comprises four recognised genera viz. *Arillastrum, Eucalyptopsis, "Stockwellia"* and *Allosyncarpia,* but surprisingly only 5 species in total, thus reflecting the wide morphological divergence of its members. So far only three species have been examined for their essential oils. The 2 species of *Eucalyptopsis* (from Papua New Guinea) remain to have their oils examined. *Allosyncarpia ternata* produces an oil containing mostly α- and β-pinene,

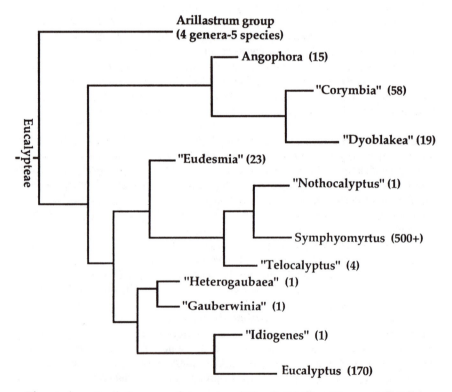

Figure 1.   A cladogram of genera within the informally erected tribe Eucalypteae following and modified from Briggs and Johnson (*1*).

1,8-cineole (1) and a smaller amount of α-terpineol (Brophy, J.J. and Boland D.J., *Flav. & Frag. J.*, in press). *Arillastrum gummiferum* (from New Caledonia) produces an oil, in poor yield, which contains approximately 80% of limonene (Brophy, J. J., unpublished data). The remaining member of this group possesses the manuscript genus name *"Stockwellia"*. This species produces an oil containing up to 90% of 2,4,6-trimethoxytoluene (*3*) which has not been found elsewhere in nature. (See Figure 2.)

Angophora Group

The eastern Australian genus *Angophora* comprises 15 species in 4 unequal sections. To date only 3 species have been examined for their oils. *A. leiocarpa* produces an oil rich in limonene (>60%) (Goldsack, R.J. and Brophy, J.J., unpublished data), *A. bakeri* produces an oil which is rich in α-pinene (Lassak, E.V. and Brophy, J.J., unpublished data), while *A. hispida* produces an oil (in 0.1% yield) which is almost entirely sesquiterpenoid in character (Lassak, E.V. and Brophy, J.J., unpublished data). The three species mentioned above come from different series within the genus and at present there are too few data.

"Corymbia". The informal genus *"Corymbia"* contains four sections within which are approximately 58 species. This grouping is commonly known as the woody-fruited bloodwoods. They occur almost exclusively in eastern Australia. For the most part, the leaves of this grouping contain little oil and they have not been investigated in detail for this reason. The species that have been examined contain high yields of α- and β-pinene, the former predominating and, low percentages of 1,8-cineole. Two Western Australian members of the group, *E. ficifolia* and *E. calophylla* also contain up to 15% of *trans,trans*-farnesol, a compound relatively rare in the eucalypts (*4*, Brophy J.J., unpublished data).

Also belonging to this group is *E. citriodora,* the lemon scented gum. This species exists in a number of chemotypes; a citronellal rich form, a form rich in citronellol and its acetate, and a hydrocarbon rich form. The latter gives a much lower oil yield than the lemon scented chemotypes (4). *E. maculata,* the spotted gum, also belongs to this group. Trees of this species from New South Wales produce an oil with a large number of sesquiterpenes (the main ones being caryophyllene and δ-cadinene and the cadinols) though the major compound is still 1,8-cineole (*5*, Brophy, J.J., unpublished data). A second chemotype, from Queensland, is reported to produce guaiol (in 40% yield) as the major compound (*5*).

"Dyoblakea". The informal genus *"Dyoblakea"* is a small grouping containing 19 species. They are commonly referred to as the paper-fruited bloodwoods. Few of these species have so far been examined for their oil, and those that have have shown poor oil yields. Three species have so far been examined from this group. Of particular interest, however, is the oil of *E. papuana,* a species of wide distribution and soon to be split into several new species. Oil from the specimens

1

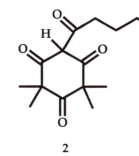

2

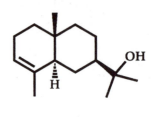

3

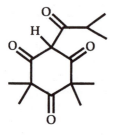

4

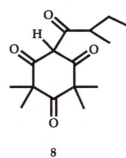

5

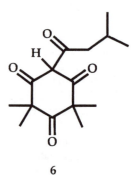

6

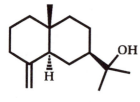

7

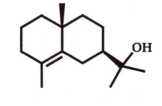

8

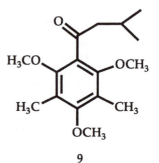

9

Structure	Common Name
1	1,8-Cineole
2	---
3	α-Eudesmol
4	β-Eudesmol
5	γ-Eudesmol
6	Leptospermone
7	Flavesone
8	Isoleptospermone
9	Torquatone

Figure 2. Chemical structures and common names. For structures with no common names, see text for names.

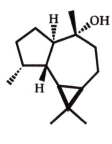

10

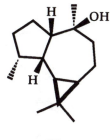

11

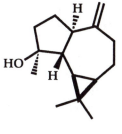

12

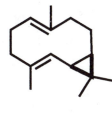

13

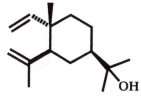

14

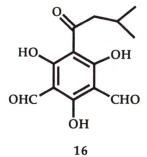

15

16

17

Structure	Common Name
10	Globulol
11	Viridiflorol
12	Spathulenol
13	Bicyclogermacrene
14	Elemol
15	Isobicyclogermacral
16	Jensenone
17	Tasmanone

Figure 2. Continued. *Continued on next page.*

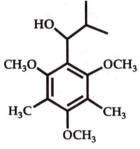

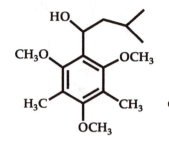

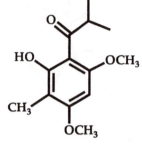

18

19

20

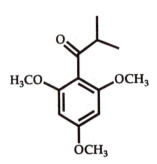

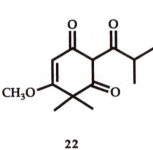

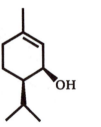

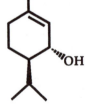

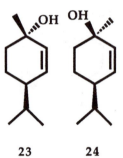

22

23   24

21

25   26

Structure	Common Name
18	---
19	---
20	Baeckeol
21	Conglomerone
22	Agglomerone
23	---
24	---
25	---
26	---
27	---

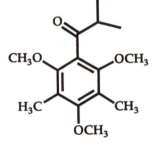

27

Figure 2. Continued.

growing at Mareeba in northern Queensland contain 25-40% of the new β-triketone (2), (Brophy, J.J. & Clarkson, J.R., unpublished data).

Eudesmia Group

"Eudesmia". The informal genus *"Eudesmia"* comprises approximately 23 species, most of which occur in Western Australia. Because of their relative geographical isolation, few species in this genus have been examined for their oils. Those that have been examined have poor oil yields (<0.5%). Of the three species so far examined *E. tetragona* contains large amounts of α-, β- and γ-eudesmols (3, 4, 5), while *E. gongylocarpa* contains significant amounts of α-pinene, 1,8-cineole and globulol (10). (Boland, D.J. and Brophy, J.J., unpublished data).

E. erythrocorys contains 60% of 1,8-cineole with small amounts of α-pinene, α-terpineol and perillyl acetate. The three species mentioned above cover two of the three sections within this informal genus and really no pattern can be discerned yet (Lassak, E.V. and Brophy, J.J., unpublished data).

"Nothocalyptus". The informal genus *"Nothocalyptus"* is monotypic, the only species being *E. microcorys,* known as tallowwood. This species, noted for its good timber, produces an oil in poor yield (~0.5%) which is rich in 1,8-cineole and has moderate amounts of the α- and β-pinenes but contains very small amounts of sesquiterpenes (6, Brophy, J.J. unpublished data).

Symphyomyrtus

By far the largest number of species of *Eucalyptus* sens. lat. belong to the informal genus *Symphyomyrtus*. This grouping contains over 500 species arranged in 9 sections, containing from 1 to in excess of 100 species. The grouping covers the whole of Australia, though certain sections are more or less confined to Western Australia and others to eastern Australia. Because of the large number of species, this group will be dealt with by sections.

Section Tingleria contains only one species, the Western Australian *E. guilfoylei,* and has not yet been examined.

Section Transversaria occurs mainly on the east coast and contains approximately 24 species, of which half have been examined. Overall, the oil yields are in the 0.5 - >2% range. They are all high in monoterpenes, and the principal component is 1,8-cineole, though in species at the lower oil yield range the principal component is α-pinene. All are low in sesquiterpenes. *E. diversicolor,* the only Western Australian member of this section contains over 30% of α-terpinyl acetate (7). *E. grandis* contains the three β-triketones, leptospermone (6), flavesone (7) and isoleptospermone (8) (total ~30%) in its essential oil (4). These compounds have not been detected in the closely related species *E. saligna* and *E. botryoides* (4, Brophy, J.J., unpublished data).

Section **Pygmaearia** contains one species, *E. pumila,* an east coast species. This species gives a high yield of oil (~2%) which is high in 1,8-cineole and low in other monoterpenes and sesquiterpenes (*4*).

Section **Aenigmataria** also contains only one species, the "sugar gum" *E. cladocalyx,* which has an isolated distribution in South Australia. This species produces small amounts of oil which contains 1,8-cineole (10-30%) and small amounts of sesquiterpenes. The oil does, however, contain over 30% benzaldehyde, presumably from hydrolysis of a cyanohydrin (*7*).

Section **Bisectaria** is the largest section within this informal genus, containing approximately 200 species, all but a few of which originate in Western Australia. Most species in this section have a very localised distribution within Western Australia. Only 25 species have at present been examined. Within this section there are 23 series. On the basis of so few species examined it is foolish to generalise too much, but having said that some trends do tend to stand out.

While oil yields vary within the section, the majority are not high, with exceptions to this being series Salmonophloiae and Sociales. These two series produce oils in >2% yield wllich are high in monoterpenes and in particular 1,8-cineole. Sesquiterpenes are either low or absent. *E. kochii,* a member of Sociales is a good candidate for 1,8-cineole production (*8*).

Certain Western Australian series, notably Gomophocephalae, Cornutae, Grossae, Strictlandianae, Salubres, and Calycogonae (all Western Australian species) produce oils in moderate yield with significant amounts of 1,8-cineole and moderate amounts of sesquiterpenes. In most cases so far examined, aromatic carbonyl compounds, most notably torquatone (**9**), have been found (Brophy, J.J. and Lassak, E.V., unpublished data).

Section **Dumaria,** once again a mostly Western Australian grouping, contains approximately 70 species in eight series of variable size. Only 7 of these species have so far been examined. The oil yields of this section are generally >1.5%, high in monoterpenes, of which the major member is 1,8-cineole. Of the 7 species so far examined, 3 have shown the presence of the aromatic ketone, torquatone (**9**). All the species contain sesquiterpenes, both hydrocarbons and alcohols of the 7.5.3 ringed series, of which the major members are aromadendrene and globulol, as well as the eudesmols (Brophy, J.J., unpublished data).

Section **Exlegaria** from the east coast of Australia contains only one species, *E. michaeliana.* The oil from this species is typical of other eucalypts in that it contains mostly 1,8-cineole (>70%) with much lesser amounts of α-terpineol, spathulenol, *trans*-pinocarveol and pinocarvone (Lassak, E.V., unpublished data). The latter two compounds may be oxidation products of α-pinene. The β-triketone, leptospermone (**6**), has also been identified in trace amounts (*9*).

Section Exsertaria is an eastern Australian grouping of approximately 60 species in three series. They are commonly known as the red gums. A considerable number of species in this grouping have been examined. Four species have been examined from Series Albae, which contains approximately 30 species. These have poor oil yields (<0.5%), are high in monoterpenes of which the major members are α- and β-pinene. 1,8-Cineole is present in only very small amounts (*4*).

Series Bancroftianae contains 7 species, all from northern New South Wales and Queensland. To date none has been examined.

Series Umbellatae contains 24 species, one third of which have been examined. Overall oil yields range from 0.5 - 1.5% with monoterpenes predominating. The nexus between low oil yield and high α- and β-pinene content also prevails in this section. Sesquiterpenes are generally more plentiful though rarely greater than 30% overall. The presence of the related compounds globulol (**10**), viridiflorol (**11**) and spathulenol (**12**) in the species so far examined is also of interest. The α-, β-, and γ-eudesmols were only found in amounts ~5% in two of the species (*4*, Brophy, J.J. unpublished data).

E. camaldulensis, the river red gum, is a member of the above series. It has the widest distribution of any of the eucalypts in Australia. There are two chemotypes in this species, one which contains greater than 70% 1,8-cineole and very low in sesquiterpenes, and a second chemotype which is rich in the sesquiterpenes globulol, viridiflorol and spathulenol (*10*). Controlled crossing experiments have shown that progeny of both chemotypes exhibit the same type of oil (rich in 1,8-cineole with no sesquiterpenes) until the trees are approximately 20 months old when those of the second chemotype apparently cease to produce 1,8-cineole and begin producing sesquiterpenes. At 25 months, the oil contains greater than 50% bicyclogermacrene (**13**), undoubtedly a precursor of globulol, viridiflorol and spathulenol (Doran, J.C. personal communication).

Section Maidenaria. This section consists of approximately 80 species in 10 unequal sections. They are basically south eastern Australian temperate species which have moderate to good oil yields (1->2%) and in most cases 1,8-cineole accounts for the greater proportion of the oil. They are unique in *Symyhyomyrtus* in having oil glands in the bark which is a taxonomic feature of the group. Sesquiterpenes can either account for 30% or more of the oil, and in these cases the compounds responsible are globulol, viridiflorol and spathulenol or a mixture of the α-, β-, and γ-eudesmols. Because most of these species are located in eastern Australian, approximately 90% of this section have been analysed. *E. globulus* and *E. smithii,* both of which are used overseas for eucalyptus oil production, belong to this section (*4*, Brophy, J.J. unpublished data).

There are exceptions to the above generalisations within this section. *E. aggregata* produces an oil that consists almost entirely of esters of phenylacetic acid, the major one (>90%) being phenylethyl phenylacetate (*4*). Its closest relative *E. rodwayi* produces an oil which is typical of most eucalyptus oils, with the major compound being 1,8-cineole and sesquiterpenes being virtually absent (*4*). Within that same series, *E. yarraensis* produces an oil in poor yield which is composed almost entirely of benzaldehyde. This undoubtedly arises from hydrolysis of a cyanohydrin (4). E. crenulata, the sole member of Series Crenulatae within this section, also produces an oil in 0.5% yield, whose major components are p-cymene and phenylethyl phenylacetate. Also present in the oil are other esters, mostly derived from phenylacetic acid, as well as methyl eudesmate and γ-terpinene (4). Within the eucalypts as a whole, E. aggregata and E. crenulata are the only two species known to date to exhibit such oils. E. macarthurii yields an oil rich in both the three eudesmols and geranyl acetate. Of the species yielding oils rich in sesquiterpenes (4, 5), E. nova-anglica stands out. It produces oils from three chemotypes, in >2% yield, all of which are rich in sesquiterpenes. These chemotypes are rich in 1.) α-, β-, and γ-eudesmols, 2.) aromadendrene, globulol, viridiflorol and spathulenol, and 3.) nerolidol. The occurrence of nerolidol is rare in the eucalyptus oils (11).

Section Adnataria. This section consists of approximately 130 species in 18 unequal series. This section, which contains the boxes and iron bark eucalypts, is found in eastern Australia though there is a higher representation from tropical Australia. Approximately one third of the species in this section have been examined. As a general rule, oil yields are in the 0.5-2% region, though the higher yields appear to be concentrated in certain series. E. polybractea, the species from which most of Australia's eucalyptus oil is produced, occurs in this section (4, Brophy, J.J., unpublished data).

Within Section Adnataria, Oliganthae is a large series containing over 40 species. Those species examined to date show moderate to poor oil yields. With the exception of E. pruinosa, which is rich in oxygenated sesquiterpenes, monoterpenes predominate in these oils, with 1,8-cineole being the principal component (4, Brophy, J.J., unpublished data).

Within this section, Series Populneae, occurs E. brownii, a species from northern Queensland whose oil yield (approximately 2%) and composition (1,8-cineole >80%) makes it a candidate for the production of eucalyptus oil in the tropics (4).

E. dawsonii, the only member of series Dawsonianae, exists in two chemotypes. One chemotype (oil yield 4%) contains over 60% of α-, β-, and γ-eudesmols as well as elemol (14) (16%) and virtually no monoterpenes. The second chemotype (oil yield 2.5%) contains isobicyclogermacral (15) (44%), trans,trans-farnesol (7%), spathulenol (9%), and another as yet unidentified sesquiterpene alcohol (14%) with the same degree of unsaturation as spathulenol as principal components (4). Isobicyclogermacral has also been found in small amounts in the unrelated species E. dwyerii (section Exertaria) (4).

Series Crebrae, which contains 25 species, includes species showing a wide variety of essential oils, even within closely related species. For example E. crebra, E. cullenii, E. staigeriana and E. jensenii all belong to the super species Crebrosae, but in these cases the oils of the four species show striking differences. E. crebra exists in two chemotypes, in one of which α- and β-pinene make up almost 60% of the oil with globulol accounting for a further 13%. The second chemotype contains 1,8-cineole (>66%) as the major compound and globulol accounts for <5%. The oil yield of this second chemotype, at 0.9% is also twice that of the first chemotype (4). *E. cullenii* produces an oil in which 1,8-cineole and p-cymene in roughly equal amounts account for approximately 75% of the oil (Brophy, J.J. and Clarkson, J.R., unpublished data). *E. staigeriana* produces a distinctive oil rich in esters (methyl geranate and geranyl acetate), neral and geranial, limonene and β-phellandrene. There is only a very small amount (0.3%) of 1,8-cineole present in this oil which is produced in approximately 3% of fresh leaf. It used to be the basis of a small essential oil industry (4, 12). The last member of this super species, *E. jensenii,* produces an oil, in approximately 0.2% yield, in which, by far the major component is the unusual fully substituted phloroglucinol derivative, jensenone (16) (4, Brophy, J.J. et al., *Phytochem.* in press). This compound bears a resemblance to the compounds reported from the unrelated species, *E. grandis* and *E. pulverulenta* (13). Like grandinone, jensenone does have some photosynthetic electron transport properties (Paton, D.M., personal communication). This group of species does highlight the difficulties in trying to generalise too greatly on oils where species grouping is determined on morphological grounds.

The remaining series within section Adnataria consist of species which produce oils of moderate to good yield and which are high in monoterpenes, the principal member being 1,8-cineole. While sesquiterpenes are relatively rare in these latter sections, when they do occur, it is the alcohols globulol, viridiflorol and α-, β-, and γ-eudesmol which predominate. *E. polybractea* occurs within this latter group.

"Telocalyptus". The informal genus *"Telocalyptus"* consists of two series which contain a total of four species. Two species have been examined. *E. deglupta,* which is one of two species to occur outside of Australia, (the other being *E. urophylla,* in the informal genus *Symphyomyrtus*), has been shown to produce an oil (in 0.1% yield) which is rich in nerolidol (60%) (14). This is one of only two reports of this compound occurring in signlficant amounts within the eucalypts. *E. raveretiana* produces an oil, also in poor yield, in which the major compound is α-pinene. Sesquiterpenes are essentially absent from this oil (Brophy, J.J., unpublished data).

Eucalyptus Group

"Gauberwinia". The informal genus *"Gauberwinia"* contains only one species, *E. curtisii.* The oil obtained from this species (in 0.1% yield) is rich in sesquiterpenes with globulol (10), viridiflorol (11) and spathulenol (12) being the main components. Significant amounts of caryophyllene and aromadendrene are also present (Boland, D.J. and Brophy, J.J., unpublished data).

"Heterogaubaea". The informal genus *"Heterogaubaea "* also contains only one species, *E. tenuipes.* This species produces an oil rich in monoterpenes with 1,8-cineole being by far the major component (Brophy, J.J. and Dunlop, P., unpublished data).

"Idiogenes". The informal genus *"Idiogenes"* contains one species, *E. cloeziana*. This species, which occurs in two distinct regions of Queensland, has been shown to exist in two chemically distinct forms. One chemotype occurs throughout the whole of the range of the species while the second chemotype occurs in only one region. The first chemotype, which yields an oil in approximately 0.5% yield consists mainly of α-pinene (78%) and β-pinene (6%). 1,8-Cineole is not present in the oil and sesquiterpenes account for less than 5% of the oil. The second chemotype produces an oil (1.7% yield) which consists almost entirely of the β-triketone tasmanone (17) (96%) (*4, 15*). While this compound has been found before in the eucalypts (see below), its occurrence in one chemotype as here is unusual. This chemotype is representative of the bulk of the trees at the site where it occurs.

Eucalyptus. This grouping represents the genus *Eucalyptus* s. str. and consists of approximately 170 species and subspecies in four unequal sections. It includes the most important commercial "timber" trees and covers the whole of continental Australia but is more heavily represented in eastern Australia. In common with the other groupings mentioned earlier, more work has been done on the eastern species than those from western Australia.

Section Islaria. This section contains one species, *E. rubiginosa,* and occurs in Queensland. The oil from this species, produced in poor yield, consists mainly of α-pinene with lesser amounts of β-pinene; 1,8-cineole being almost absent (Brophy, J.J., unpublished data).

Section Hesperia. This section consists of approximately 23 Western Australian species, of which only 4 have been examined to date. These species all show oil yields in the range 0.5-1.5% and produce oils that contain mostly 1,8-cineole. Sesquiterpenes, while being present are not plentiful. *E. calcicola,* while producing an oil rich in monoterpenes also contains the aromatic alcohols (18), (19), and baeckeol (20) (Boland, D.J. and Brophy, J.J., unpublished data).
E. brevistylis produces an oil, in low yield, which is rich in the sesquiterpenes, in particular caryophyllene, humulene and δ-cadinene and a large number of mostly unidentified sesquiterpene alcohols. It also contains 20% of the aromatic ketone conglomerone (21); 1,8-cineole is not present in the oil (Brophy, J.J., unpublished data).
Section Renantheria. The main section of this genus is Section Renantheria. This contains approximately 120 species, the great majority of which come from eastern Australia and have been examined. The trends which emerge in this section are that oil yields are generally moderate to good (0.5->2%). When the oil yield is good, 1,8-cineole is the major component. Sesquiterpenes occur in all oils and α-, β-, and γ-eudesmols are nearly always the major sesquiterpenes (usually to the exclusion of the globulol, viridiflorol and spathulenol). Within this

section there is a group of closely related species which are rich in β-triketones and also a group of series relatively rich in *cis*- and *trans*-p-menth-2-en-1-ol together with *cis*- and *trans*-piperitol, as well as piperitone. Within the whole of the tribe Eucalypteae, this group of compounds occurs, in any quantity, only within this group of species.

In most series the oil contents are similar, as mentioned above. There are exceptions, however. The closely related species, *E. macrorhyncha, E. youmanii, E. williamsiana, E. stannicola, E. subtillior*, all produce oils which are rich in 1,8-cineole and α-, β-, and γ-eudesmol (even to the extent that the oils usually crystallise) (*16*, Boland, D.J. and Brophy, J.J., unpublished data). A further member of this group, *E. cannonii*, produces an oil virtually devoid of monoterpenes, containing large amounts of globulol, viridiflorol and spathulenol and the two trimethoxy aromatic compounds torquatone (**9**) and its corresponding alcohol (**19**) (Brophy, J.J., unpublished data).

Within this section, the subseries Capitellatosae contains *E. agglomerata, E. capitellata, E. camfieldii* and *E. bensonii*, all of which contain the β-triketones agglomerone (**22**) and/or tasmanone (**17**) (*4*, Brophy, J.J., unpublished data), while the closely related species *E. conglomerata* contains conglomerone (**21**) (*17*) and *E. blaxlandii* oil contains α-pinene and limonene as its major compounds but no aromatic components.

Following this subseries comes a group of over 70 species from seven series practically all of which contain *cis*- and *trans*-p-menth-2-en-1ol (**23**) (**24**) together with *cis*- and *trans*-piperitol (**25**), (**26**). These undoubtedly arise from a common precursor (*18*). The majority of the species also contain significant amounts of piperitone. Some regional differences can also be found within species. *E. delegatensis*, a species that occurs in both south eastern mainland Australia and the island state of Tasmania, gives an oil from mainland Australia contains 4-phenylbutan-2-one while that from the Tasmanian trees contains a statistically significant smaller amount of this compound (*19*). *E. olida* does not conform to the pattern mentioned above. This species, which comes from northern New South Wales, produces an oil containing up to 95% of *trans*- methyl cinnamate; the two p-menth-2-en-1-ols (**23**) and (**24**) and the two piperitols (**25**) and (**26**) are not found in this oil (*20*). Other species originating from Tasmania are not noticeably different from the mainland relatives, though *E. pulchella* contains 20% of the trimethoxy aromatic ketone (**27**) and its corresponding alcohol (**18**) as well as a small quantity of torquatone (**9**), *E. risdonii* contains small amounts of baeckeol (**20**) (Brophy, J.J., unpublished data) and *E. coccifera* contains 2% of *trans*-methyl cinnamate (Brophy, J.J., unpublished data).

Conclusions

With only approximately 30% of the known species examined and even then with a very uneven sampling, it is not justifiable in drawing too many hard and fast conclusions. A few trends have, however, emerged especially in groups for which the majority of species have been examined. The most significant trend is in the informal genus *Eucalyptus* sens. strict., section Renantheria. A large group of species in this section contain the *cis-* and *trans*-p-menth-2-en-1-ol and *cis-* and *trans*-piperitol, as well as in a significant number of species, piperitone. These compounds do not occur elsewhere in the eucalypts. Also within this section, is a group of species whose oils consist almost entirely of β-triketones, but these compounds are not exclusive to the section.

There seem to be more regional differences that cut across species boundaries, e.g. there seems to be a much greater chance of aromatic compounds occurring in species from Western Australia than those from the eastern side of the continent. This may reflect the past history of the continent which, in the Cretaceous period, possessed a vast inland sea in what is now central Australia (*21*). Species from Western Australia are quite numerous; they often occur in small isolated and sometimes inaccessible areas. A lot of work still has to be done before a better chemical picture of the species from this area can be constructed.

Acknowledgements

We wish to thank M. I. H. Brooker, J. C. Doran, C. J. Fookes, R. J. Goldsack, A. P. N. House, D. A. Kleinig, and E. V. Lassak and who have helped with the overall project. This project has been supported by the Australian Centre for International Agricultural Research.

Literature Cited

1.   Johnson, L. A. S.; Briggs, B. G. Myrtaceae. In *Flowering Plants of Australia;* Morley, B. D.; Toelken, H. R., Eds.; Rigby, N. S. W., **1983**, pp 175-185.

2.   Pryor, L. D.; Johnson, L. A. S. *A Classification of the Eucalypts;* Australian National University Press, Canberra, **1971**.

3.   Brophy, J. J.; Fookes, C. J. R.; House, A. P. N. H. *Phytochem.*, **1992**, *31*, 324-325.

4.   Brophy, J. J.; House, A. P. N. H.; Boland, D. J.; Lassak, E. V. In *Eucalyptus Leaf Oils - Use, Chemistry, Distillation and Marketing;* Boland, D. J.; Brophy, J. J.; House, A. P. N. H., Eds.; Inkata Press, Melbourne, Victoria, **1991**, pp 29-155.

5.   Lassak, E. V.; Brophy, J. J.; Boland, D. J. In *Eucalyptus Leaf Oils Use, Chemistry, Distillation and Marketing;* Boland, D. J.; Brophy, J. J.; House, A. P. N. H., Eds.; Inkata Press, Melbourne, Victoria, **1991**, pp 157-183.

6.   Jones, T. G. H.; Lahey, F. N. *Proc. Roy. Soc. Qld.* **1938**, *50*, 43-45.

7.   Dellacassa, E.; Menendes, P.; Moyna, P.; Soler, E. *Flav. & Frag. J.,* **1990**, *5*, 91-95.

8.   Brooker, M. I. H.; Barton, A. F. M.; Rockel, B.A.; Tjandra, J. *Aust. J. Bot.*, **1988**, *36*, 119-129.

9.   Hellyer, R. O. *Aust. J. Chem.*, **1968**, *21*, 2825-2828.

10.   Doran, J. C.; Brophy, J. J. *New Forests,* **1990**, *4*, 25-46.

11. Brophy, J. J.; Lassak, E. V.; Boland, D. J. *J. Ess. Oil Res.*, **1992**, *4*, 29-32.

12. Porich, F.; Farnow, H.; Winkler, H. *Dragoco Report*, (English Edn.) **1965**, *9*, 175-177.

13. Yoshida, S.; Asami, T.; Kawano, T.; Yoneyama, K.; Crow, W. D.; Paton, D. M.; Takahashi, N. *Phytochem.*, **1988**, *27*, 1943-1946.

14. Martínez, M. M.; Guimerás, J. L. P.; Hernández, J. M.; Zayas, J. R. P.; Días, M. J. Q.; Montejo, L. *Rev. Cub. Farm.*, **1986**, *20*, 159-168.

15. Brophy, J. J.; Boland, D. J. *J. Ess. Oil Res.*, **1990**, *2*, 87-90.

16. Brophy, J. J.; Lassak, E. V.; Win, S.; Toia, R. F. *J. Sci. Soc. Thailand*, **1082**, *8*, 137-145.

17. Lahey, F. N.; Jones, T. G. H. *Proc. Roy. Soc. Qld*, **1939**, *51*, 10-13.

18. Lassak, E. V.; Southwell, I. A. *Phytochem.*, **1982**, *21*, 2257-2261.

19. Boland, D. J.; Brophy, J. J.; Flynn, T. M.; Lassak, E. V. *Phytochem.*, **1982**, *21*, 2467-2469.

20. Curtis, A.; Southwell, I. A.; Stiff, I. A., *J. Ess. Oil Res.*, **1990**, *2*, 105-110.

21. Gould, R. E. In *Early Australian Vegetation History: Evidence from Late Palaeozoic and Mesozoic Plant Megafossils;* Smith, J. M. B., Ed.; in *A History of Australian Vegetation;* McGraw-Hill: Sydney, .S.W., **1982**, pp 32-43.

RECEIVED August 19, 1992

Chapter 8

Lemon and Lime Citrus Essential Oils

Analysis and Organoleptic Evaluation

Theresa S. Chamblee and Benjamin C. Clark, Jr.

Corporate Research and Development Department, Coca-Cola Company, P.O. Drawer 1734, Atlanta, GA 30301

The techniques used in this laboratory to obtain accurate qualitative and quantitative analyses of lemon and lime essential oils are reviewed. This quantitative database has helped to clarify the differences among Sicilian and California lemon peel oils, Mexican lime peel oil, and also to explain the chemical changes that take place during production of distilled lime oil. The HPLC-separated oxygenated fractions of Sicilian lemon peel oil have been evaluated organoleptically by a panel of experts using a capillary GC sniff port. A relatively quick, convenient method was used to pinpoint important contributors to lemon aroma. Although the data is based on a small number of opinions, the results agree well with earlier literature and expand the knowledge of lemon flavor. Relative intensity factors were assigned to lemon oil oxygenated constituents which were then ranked according to their importance to lemon aroma. Many unidentified trace constituents were also evaluated and ranked.

Lemon and lime essential oils, used principally as flavors and fragrances, are of considerable commercial importance. The U.S. is a leader in lemon oil production, accounting for about 25% of the world production in the 1980's (1). In addition, large quantities of lime and lemon oil are imported into the United States making them second and third, respectively, in terms of import dollars spent on essential oils (2). Considering the economic importance of these oils, there is a need to study them in detail. Three areas of information are required in order to understand the complex nature of essential oils: qualitative (which constituents are present), quantitative (how much of each constituent is present), and organoleptic (which constituents are most important to the flavor and end use of the essential oil). The latter can be the most important but it is often the least understood. In order to better evaluate an essential oil and its usefulness, all three areas should be explored. This paper will review some of the techniques that have been helpful to us in developing a qualitative and

0097–6156/93/0525–0088$06.00/0

quantitative database for lemon and lime oils, and it will also present some details of an approach used to evaluate a lemon oil organoleptically.

Qualitative Analysis: LC and GC-MS

The most widely used methods for separation, component identification, and quantitative analysis of essential oils are GC and GC-MS. As is well known, even capillary GC and GC-MS are inadequate to deal completely with the many coeluting and partially resolved components found in these samples. Normal phase LC, both open column and especially HPLC, are useful prefractionation tools that can greatly facilitate the GC-MS identification of essential oils. By combining LC and GC, one is able to take advantage of multiple separation mechanisms. Normal phase LC separates on the basis of polarity and affinity of the compounds for silica while GC adds the effects of vapor pressure and temperature. This combination results in very powerful separations.

In the early 1980's we developed an HPLC method for the prefractionation of citrus essential oils and terpene mixtures (*3-5*). The essential oils are initially separated by open column chromatography into two fractions, first by eluting with hexane to remove the hydrocarbons, and then with CH_2Cl_2 to remove the oxygenated constituents. Each fraction is then further separated by normal phase silica HPLC into several sub-fractions that can be analyzed by GC-MS or other techniques. One last fraction (OC-3) was obtained by washing the open silica gel column with a 10:90 mixture of methanol-methylene chloride after most of the oxygenated fraction was removed with CH_2Cl_2.

The HPLC system that is used consists of a Waters M6000 pump with 3 columns connected in tandem: Whatman Partisil 5 µm, 10 µm, and a Waters silica radial compression column. Refractive index detection is used. One advantage of normal phase liquid chromatography is the use of a nonaqueous mobile phase which allows for very easy sample work-up with no extractions necessary before further analysis of the collected samples. A series of mobile phases was tried with the best overall separations being achieved with a ternary solvent mixture of 8% ethyl acetate in a 50:50 mixture of hexane/methylene chloride.

Normal phase HPLC provides a good class separation of the citrus essential oils. Elution from the silica column occurs in order of increasing polarity. The monoterpene hydrocarbons elute first, followed by the sesquiterpenes, esters, aldehydes, and finally alcohols. Figure 1 shows an HPLC of the oxygenated fraction of Sicilian lemon oil. The largest peak (fractions 6 and 7) in the chromatogram is principally citral which comprises close to 60% of the total oxygenated fraction. By fractionating this sample using HPLC, it was possible to greatly simplify the chromatograms for GC-MS, as shown in Figure 2. The chromatogram in Figure 2A shows the Sicilian lemon oil oxygenated fraction, and GC's of two representative sub-fractions (#8 and #12) collected from the HPLC are shown in 2B and 2C respectively. It is easy to see that GC-MS of these fractions will be simpler than that of the starting material. Also there is the added advantage that certain compounds which coelute in GC are separated by HPLC.

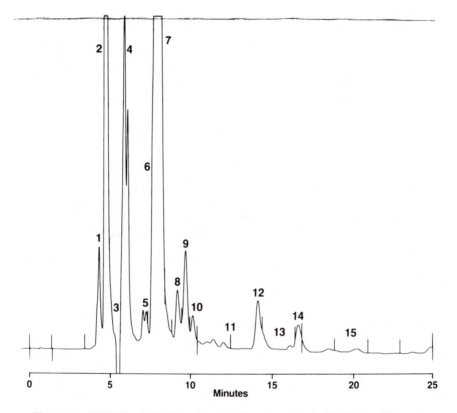

Figure 1.   HPLC of Sicilian lemon oil, oxygenated fraction, with a few representative compounds listed: fractions 1 and 2, solvent and trace monoterpene oxygenates; fraction 3, neryl acetate, nonanal; fraction 4, geranyl acetate, octanal; fraction 5, neral; fraction 6, neral, geranial; fraction 7, geranial; fraction 8, terpinen-4-ol; fraction 9, linalool; fraction 10, campherenol; fraction 11, unidentified; fraction 12, *t*-sabinene hydrate, borneol; fraction 13, citronellol; fraction 14, α-terpineol; fraction 15, geraniol.

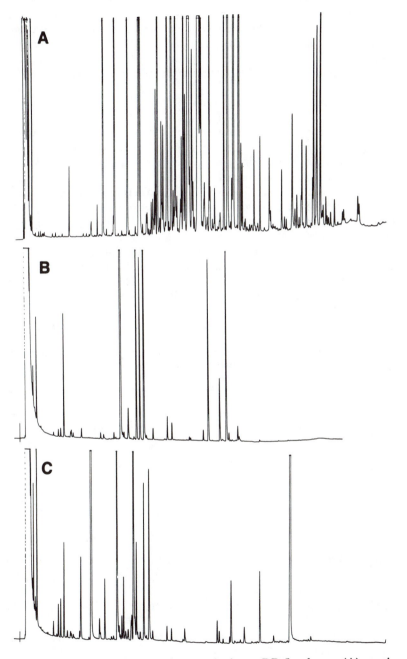

Figure 2.   Capillary GC separation employing a DB-5 column: (A) starting material, Sicilian lemon oil oxygenates; (B) HPLC fraction #8 (see Figure 1); (C) HPLC fraction #12.

The hydrocarbon constituents are also well separated by normal phase HPLC (6). The optimum mobile phase in this case is 100% hexane. Monocyclic and bicyclic hydrocarbons are reasonably well separated and the sesquiterpene hydrocarbons are well resolved. This was especially useful when identifying a flavor impact compound of expressed lime oil, germacrene B. The final identification of this compound was made possible with a sample isolated and purified by HPLC (6).

Quantitative Analysis: Lemon and Lime

The identification of additional constituents in lemon and lime oils has allowed us to improve and expand our quantitative database for these essential oils. A good quantitative database, obtained using response factors and internal standards, is useful for many applications. Unfortunately, as Shaw has pointed out (7), much of the analytical data available in the literature for citrus oils is based on relative peak area percents without the use of response factors and internal standards. Our quantitative work has provided absolute weight percent values for 52 constituents in Sicilian and California lemon oils (8), 50 constituents in Mexican expressed lime oil, and 42 constituents in Mexican distilled lime oil (9). These analyses represent the most complete published quantification of these oils at present. A summary by compound class of quantitative data for lemon and lime oils is shown in Table I. A complete compilation of the quantitative data for individual constituents is available (8, 9).

Table I. Quantitative Volatiles Analysis: Summary by Compound Class
Absolute weight percent[a]

	Sicilian Lemon[b]	California Lemon[b]	Expressed Lime[c]	Distilled Lime[c]
Hydrocarbons, C(10)				
Bicyclic	14.62	15.17	25.36	3.31
Monocyclic	76.44	76.32	53.59	72.13
Total	91.06	91.49	78.95	75.44
Hydrocarbons, C(15)	1.16	1.24	5.70	3.48
Aldehydes	3.78	2.32	5.00	0.48
Alcohols	0.61	0.61	1.05	13.41
Esters	0.96	1.18	0.44	0.15
Ketones, Ethers	0.03	0.03	0.04	3.96
Unidentified	0.20	0.29	0.76	2.38
Total GC Volatiles	97.80	97.16	91.94	99.30

[a] Corrected for GC flame detector response and using tetradecane as internal standard. [b] From ref. 8. [c] From ref. 9.

Quantitative data have proved very useful for explaining differences among the lemon and lime oils. For example, the data in Table I show that the Sicilian lemon oil is rich in aldehydes but slightly lower in esters compared to the California lemon oil. Expressed lime oil, which is obtained directly from the peel of the fruit, has a relatively high level of aldehydes, especially citral, and high levels of bicyclic hydrocarbons compared to lemon oil. In addition, expressed lime oil also contains significant amounts of nonvolatile material. Differences between the expressed and distilled lime oils are also evident. Distilled lime oil is rich in alcohols and low in aldehydes and bicyclic hydrocarbons compared to the expressed lime oil. The quantitative data can be used to understand the changes that take place during the production of distilled lime oil, which is actually a reaction flavor made by acid-catalyzed reaction of crushed lime fruit. The acid-catalyzed reactions of the bicyclic hydrocarbons that occur during production of distilled lime oil result in an increase of terpene alcohols, especially α-terpineol. The terpene aldehydes are also almost completely lost during production. These reactions have been summarized in detail (9).

Organoleptic Analysis

In general, the more extensive the identification of constituents in these oils, the more accurate the quantitative analysis can be. Quantitative data provide a basis for understanding the correlation between quality and composition. The third component required for this correlation is organoleptic analysis. Organoleptic analysis is very complex and encompasses many variables that are hard to control, yet it has received much attention over the years. This is not an especially standardized area of study and many methods have been tried. One well known approach focuses on the use of odor thresholds which are used to calculate an aroma value or odor unit (10, 11). Odor units are ratios of the concentration of a particular constituent in a food or flavor to the calculated threshold of the component. They have been used extensively by Guadagni, Buttery, Teranishi and coworkers (12). Threshold values have been calculated for many individual standards in water. Pure standards of the compounds of interest are necessary and sometimes difficult to obtain. Nevertheless, this technique has been used to identify important flavor constituents in many vegetables and fruits. Sugisawa and coworkers have used a similar approach to study odor quality of orange (13). Another interesting flavor evaluation was performed by Casimir and Whitfield on passion fruit using flavor impact values (FIV) which were correlated to taste panel scores (14).

Other approaches for organoleptic evaluation use a capillary sniff port to help calculate thresholds. For example, Drawert and Christoph used aroma values to rank the importance of 13 constituents in the headspaces of lemon oil, lemon peel, and lemon oil emulsion (15, 16). These aroma values represent the ratio of the headspace concentrations of the constituent to the thresholds determined using a sniff port. Grosch and coworkers also investigated lemon with a GC sniff port method using flavor dilution values (17) for threshold calculation. Flavor dilution values were calculated for twelve constituents in fresh lemon oil (18). The flavor dilution technique has also been applied to aged and oxidized lemon oil samples (19) and many other flavors. Acree's "charm" analysis is a similar capillary sniff port method. It involves calculation of charm values which are related to thresholds and odor units (20). The method involves the capillary sniff port analysis of a series of dilutions of the sample of interest and can be applied to components of unknown identity or concentration (21, 22).

Threshold information is very useful in flavor analysis; however, the approach does have some limitations. Threshold determination is time consuming and generally requires a large, trained panel. To be most effective, thresholds should be determined in a medium that is appropriate for the sample of interest (23, 24). There is often large variability in threshold measurements reported from laboratory to laboratory (25). This may be due to the fact that pure standards are often difficult to obtain. In the case of unidentified constituents and even some known compounds, pure standards are not available. The capillary sniff port methods avoid some of these limitations. Unknown constituents can be evaluated; but, the possibility of coeluting trace constituents still exists. Analyzing several dilutions of a complex essential oil by capillary sniff port can be very time consuming.

Another issue relating to thresholds is the possibility of the additive effects of sub-threshold constituents in a mixture. There are several reports in the literature describing the additivity of flavor compounds to produce an aroma even when they are present below their individual thresholds (24, 26-30). These reports give strong evidence that compounds present in a mixture below their threshold should not be ignored.

Capillary Sniff Port Approach for Lemon Oxygenates

The above issues and limitations have led us to a somewhat simpler, quicker approach for organoleptic analysis that uses a capillary GC sniff port and does not rely on threshold determination. A similar method for generation of initial intensity data was used by von Sydow et al. to study bilberry flavor (31). The objective of our analysis was to determine the relative importance of constituents, particularly trace unknowns, present in the oxygenated fraction of Sicilian lemon oil. Efforts were concentrated on the oxygenates because a preliminary sniff port test of the hydrocarbon fraction suggested that it is much less important to the lemon aroma. Citral is a well known lemon flavor impact compound, but it would be useful to determine the significance of the other constituents as well. Over 140 compounds have been reported to be present in lemon peel oil (32). Unfortunately, some of these identifications were reported with little or no supporting analytical data. We have confirmed the presence of over 60 compounds in Sicilian lemon oil (8 and (Clark, Jr., B. C.; Chamblee, T. S.; The Coca-Cola Company, unpublished data)). Our GC-MS analysis of the HPLC fractions for the lemon oil oxygenates indicated ≈150 compounds that were not identified or confirmed by us. Most of these are trace constituents because only approximately 0.20 weight % of lemon oxygenates remain unidentified (8). Trace constituents have been found to be important in other essential oils (33). Since identification of trace constituents can be difficult, it would be especially helpful to know which of the unidentified compounds are organoleptically important. This capillary sniff port approach does not attempt to provide a rigorous organoleptic evaluation of the oil but rather is designed to be used as a guideline for pinpointing important aroma constituents.

Experimental.   A panel of five people, chosen especially for their familiarity with citrus flavors, was used. The panelists knew they were smelling lemon fractions and were instructed to note any odors that were especially lemony or citrus-like. However, during the study they were not aware of which fraction they were smelling at a particular time or which compounds might be present in the fraction. In two sessions, the lemon fractions were first evaluated by blotter sniff test. The panelists smelled 15 fractions collected from the HPLC (Figure 1) and one fraction (OC-3) of additional polar constituents eluted from the open silica gel column with methanol-methylene chloride (10:90, v/v) after the methylene chloride fraction. A GC of the open column fraction showed that it contained trace amounts of most of the oxygenated constituents. The panelists were asked to describe each fraction with respect to its importance to lemon flavor. The fractions containing most of the citral (fractions #6 and 7) were described as very lemon-like, as expected. However, interesting comments, including additional citrus notes were recorded for several of the other fractions as well. These results were used to choose six fractions for further analysis by capillary sniff port: #2, 3, 4, 11, 12, and OC-3.

The capillary sniff port set-up is installed in a Varian 3700 GC and it has the advantage of being very simple. An SGE fused silica capillary outlet splitter is attached at the end of the capillary column. The splitter consists of two capillaries of different diameter yielding a split ratio of 10:1. The capillaries of the splitter are connected so that ten parts of the GC effluent go to the sniff port and one part goes to the FID detector. The sniff port, which is available from Varian, consists of a heated tube inside an empty thermal conductivity detector block through which the larger capillary is threaded to the atmosphere. The detector block has an independent temperature control set at 240°C. A stream of humidified air has been added at the sniff port outlet to prevent nose dry-out, although this does not seem to be much of a problem with this set-up, probably because the panelist is only smelling the flow coming from the capillary column, $\approx$1-5 ml/min. A megabore (0.5mm i.d.) DB-5 capillary column was used. There was an excellent correlation of retention time between the FID and sniff port outlet, no band broadening, and no carry over from peak to peak even with very large and intense peaks. The typical sample concentration injected was generally in the ng/peak or smaller range. However, it is well known that peak size is not necessarily indicative of odor intensity.

Prefractionation of essential oils is required for sniffing because a whole oil chromatogram will have many overlapping peaks. Also, the panelist can be overwhelmed by too many odors if the column is overloaded. An attempt was made to inject just enough sample on the GC so that the large peaks would not be too overpowering. Therefore, in a relatively dilute run a strong odor would denote a fairly potent compound, especially if the peak was very small.

The sniffers were instructed to smell the GC effluent, describe the aroma in general flavor terms, and make an estimate of the intensity for each odor detected. The panelists did not pay particular attention to whether a peak was eluting or not. A second person recorded aroma descriptions and intensity evaluations directly on the chromatogram at the retention time where they were noted. Each session lasted ≈60 minutes. Figure 3 shows a sample GC sniff port chromatogram. Intensity evaluations were converted to a numerical scale: 1= very weak, 2 = weak, 3 = moderate, 4 = strong, 5 = very strong. The HPLC exhibits tailing of some of the larger components (Figure 1), resulting in some carry over of major constituents from fraction to fraction. Therefore, several constituents are present in more than one fraction, especially citral which comprises ≈60% of the total oxygenated fraction. This turned out to be an advantage for sniff port analysis because it allowed for smelling of many of the constituents at different concentrations without going through dilutions of the sample.

Results. Correlation of relative retention time data with previous GC-MS results (8) allowed for identification of the known major constituents of lemon in the fractions that were analyzed by sniff port. A few identifications remain tentative(t). An analysis of the comments showed that the following identified peaks were described as citrus or lemon-like by more than one panelist: citronellal, citronellol, decanal, dodecanal, geranial, geraniol, linalool, neral, nonanal, octanal, perilla aldehyde(t), α-terpineol, and undecanal. The following compounds were judged to make little or no contribution to the aroma: α-bisabolol, 1,8-cineole(t), citronellyl acetate, geranyl propionate, geranyl acetate, *cis*- and *trans*-limonene oxide, methyl geranoate, neryl propionate, 2,3-dimethyl-3-(4-methyl-3-pentenyl)-2-norbornanol(t), and *trans*-sabinene hydrate. Finally, these compounds were described as very potent yet not lemon-like or citrus-like: borneol (very strong musty), linalool (moderately strong floral), terpinen-4-ol (strong earthy, musty), and undecyl acetate (strong unpleasant). Linalool was described as lemon-like and citrus by some sniffers and floral by others. These results agree with and expand evaluations of lemon flavor found in the literature (15, 16, 18, 19). 1,8-Cineole may be expected to have a greater effect than noted here. Since we used a very small panel and the identification was based on retention time correlation with previous GC-MS data, the low contribution of 1,8-cineole should be viewed as tentative. There is some indication that 1,8-cineole is subject to specific anosmia (34).

Further analysis of the sniff port chromatograms revealed 39 unidentified peaks that were described as lemony, citrus-like or having some other strong aroma. Many of these peaks are very small as indicated in Figure 3. The numbered peaks in this chromatogram are a sample of the constituents that have not been identified by us but were recognized as being potentially important to the lemon flavor.

Ranking Importance to Lemon Flavor. In an attempt to rank the relative importance of the known and unknown constituents found in the Sicilian lemon oil oxygenated fraction, the following analysis was carried out. A relative intensity factor was calculated for each peak by dividing the intensity scale number by the peak height (in cm) and multiplying by 100. Therefore, a small peak that is very potent will have a larger relative intensity factor than a large peak with moderate intensity. The relative intensity factors for the major identified peaks in the oxygenated fraction of lemon oil are shown in Table II. As mentioned previously, some constituents were present in more than one fraction. Consequently, the relative

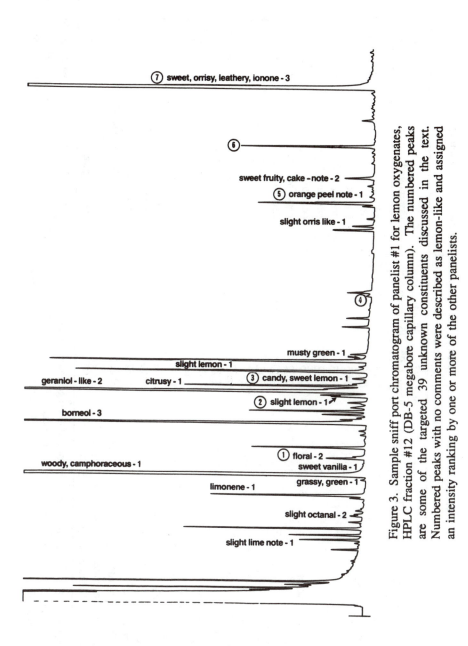

Figure 3. Sample sniff port chromatogram of panelist #1 for lemon oxygenates, HPLC fraction #12 (DB-5 megabore capillary column). The numbered peaks are some of the targeted 39 unknown constituents discussed in the text. Numbered peaks with no comments were described as lemon-like and assigned an intensity ranking by one or more of the other panelists.

Table II. Average Relative Intensity Factors and Estimated Significance to Lemon for the Identified Oxygenates of Sicilian Lemon

Compound	Average Intensity Factor	Wt % in Whole Oil[a]	Significance to Lemon[b,c]	Compound	Significance to Lemon
linalool	26.7	0.18	48.1	geranial	164.0
borneol	13.7	0.009	1.2	neral	88.2
decanal	13.4	0.05	6.7	linalool	48.1
perillaldehyde	12.8	0.03	3.8	nonanal	10.7
undecanal	11.7	0.03	3.5	citronellal	9.4
citronellol/nerol	11.3	0.03	3.4	octanal	9.4
geraniol	10.8	0.02	2.2	neryl acetate[d]	7.3
octanal	9.4	0.10	9.4	decanal	6.7
nonanal	8.9	0.12	10.7	α-terpineol	4.0
geranial	8.0	2.05	164.0	perilla aldehyde	3.8
citronellal	7.2	0.13	9.4	undecanal	3.5
neral	7.0	1.26	88.2	citronellol/nerol	3.4
terpinen-4-ol	6.7	0.05	3.3	terpinen-4-ol	3.3
undecyl acetate	6.4	0.006	0.4	geraniol	2.2
dodecanal	4.4	0.008	0.4	borneol	1.2
campherenol	4.0	0.03	1.2	campherenol	1.2
α-terpineol	2.1	0.19	4.0	undecyl acetate	0.4
neryl acetate[d]	1.4	0.52	7.3	dodecanal	0.4
methyl geranoate	1.1	0.005	0.05	t-sabinene hydrate/octanol	0.3
citronellyl acetate	0.8	0.03	0.2	citronellyl acetate	0.2
t-sabinene hydrate/octanol	0.6	0.05	0.3	methyl geranoate	0.05

a From ref. 8.   b Significance to lemon = (Average Intensity Factor X Wt% in Whole Oil) X 10   c The following compounds were judged to make no contribution to the lemon aroma using this method: 1-8,cineole, geranyl propionate, geranyl acetate, cis- and trans-limonene oxide, 2-3-dimethyl-3-(4-methyl-3-pentenyl)-2-norbornanol(t).   d See text for discussion of result.

intensity factors in Table II represent averages of multiple determinations derived from different fractions and panelists. For example, nonanal was evaluated in fractions #3, 4, and OC-3. A total of 11 aroma evaluations were used to give the average relative intensity value of 8.9 shown in Table II. As expected, when dealing with organoleptic analysis, especially with a small panel, a range of intensity values was recorded. A typical example is the average intensity value of 13.4 for decanal which was calculated from individual evaluations ranging from 7.5 to 27.8. In addition, peak height can be difficult to measure for extremely small peaks. Therefore the values in Table II should be viewed as estimates or relative evaluations rather than rigorously defined parameters of intensity. The relative intensity is a measure of the relative potency of a compound, and it may relate inversely to threshold values. For example, a large intensity factor probably indicates a compound with low threshold.

The intensity factors were used to rank the compounds according to their importance to the lemon flavor. An estimated "significance to lemon" factor was calculated as the intensity factor times concentration of the constituent in lemon oil times 10. The identified constituents are listed in Table II both in order of average intensity and estimated significance to lemon. As expected, citral, other aldehydes, and linalool rank high in importance to lemon oil aroma. Neryl acetate seems high based on results for other acetates and may be a misidentification.

Relative intensity factors and estimated significance to lemon values were calculated for the unidentified constituents also. Based on the estimate that there is ≈0.2 % unidentified GC volatile oxygenated material in Sicilian lemon oil (8) and approximately 150 constituents remain unidentified, an average concentration of 0.001% was assigned to each of the unknown trace constituents. Some compounds may actually be present at a much lower or in a few cases five times or more higher concentration than 0.001%; therefore, this value represents a crude but reasonable estimate for concentration of the unknowns. The significance to lemon ranking for the 39 targeted unidentified peaks is shown in Table III. Unlike the major identified constituents, the smaller unidentified peaks were usually not present in more than one or two fractions; therefore, the values in Table III are averages based on fewer individual evaluations compared to the identified constituents in Table II. Nevertheless, the results indicate that there are several unidentified peaks with relatively high significance to lemon values. As expected, several of these peaks have large intensity factors, suggesting that they are compounds with relatively low thresholds. Unidentified peaks in the selected lemon fractions, which have an estimated significance above 5, are now being targeted for further identification, potentially simplifying the task of understanding lemon flavor.

Despite the recognized limitations of this method, our results are in good agreement with the lemon constituents evaluated in the literature, and they significantly expand the data for lemon organoleptic analysis. Schieberle and Grosch rank the importance of twelve constituents (hydrocarbons and oxygenates) in fresh lemon oil (18). The most potent aroma compounds were found to be neral, geranial and linalool, followed about one order of magnitude lower by myrcene, limonene, γ-terpinene, nonanal, 2(E)-nonenal, decanal, octanal, and citronellal. Drawert and Christoph ranked hydrocarbons and oxygenated constituents in lemon headspace for

Table III. Estimated Significance to Lemon: Unidentified Constituents

Lemon Fraction[a]	Peak #	Average Intensity Factor	Estimated Significance to Lemon
Fraction #2	1	3	<0.1
	2	313	3.1
	3	230	2.3
	4	1000	10.0
	5	2222	22.2
	6	423	4.2
	7	12	0.1
Fraction #3	1	2500	25.0
	2	1405	14.0
	3	875	8.7
	4	4000	40.0
Fraction #4	1	1625	16.2
	2	360	3.6
	3	26	0.3
	4	625	6.2
	5	86	0.9
	6	126	1.3
	7	500	5.0
Fraction #11	1	411	4.1
	2	150	1.5
	3	179	1.8
	4	250	2.5
	5	8	0.1
Fraction #12	1	214	2.1
	2	164	1.6
	3	361	3.6
	4	500	5.0
	5	1667	16.7
	6	93	0.9
	7	12	0.1
Fraction #OC-3	1	300	3.0
	2	90	0.9
	3	20	0.2
	4	64	0.6
	5	1000	10.0
	6	16	0.2
	7	4000	40.0
	8	1333	13.3
	9	66	0.7

[a] See Experimental.

a total of 13 compounds: limonene, myrcene, octanal, γ-terpinene, neral, geranial, 1,8-cineole, linalool, ethyl hexanoate, nonanal, geraniol, octanol, and nonanol (*15*). This report suggests that limonene is more important than citral to fresh lemon aroma. The authors recognized the limitation of their technique because it did not take into account or distinguish the quality of an odor. Nevertheless, five of our top six oxygenated constituents were also cited as important in this study as well. The total number of oxygenated constituents ranked for importance in lemon oil has now been increased to 28 known compounds and 39 unidentified constituents.

Conclusion

1. The qualitative and quantitative database for lemon and lime oils has been expanded and improved. Analytical methods have been developed that can be applied to a variety of essential oils.
2. The capillary sniff port procedure used to evaluate the lemon oxygenates provides a relatively fast and simple approach for pinpointing the important aroma constituents of complex essential oils. It does not require the use of threshold data and has been shown to generally agree with and expand more complicated analyses reported in the literature.
3. The organoleptic evaluation of lemon oxygenates demonstrates that several of the unidentified trace constituents in Sicilian lemon oil are potentially important to the lemon flavor. As a result of this analysis, we will be able to target further identification efforts on constituents that are most important to lemon flavor.

Acknowledgments

We thank G. A. Iacobucci, T. Radford, R. E. Biggers, J. K. Johnson, A. S. Olansky, A. K. Ray, and N. R. White, all of these laboratories, for helpful discussion and participation in odor evaluation.

Literature Cited

1. Lawrence, B. M. *Perfum. Flavor.* **1985**, *10(5)*, 1, 3-10, 12-16.
2. Pisano, R. C. *Perfum. Flavor.* **1986**, *11(5)*, 35-41.
3. Jones, B. B.; Clark, Jr., B. C.; Iacobucci, G. A. *J. Chromatog.* **1979**, *178*, 575-578.
4. Jones, B. B.; Clark, Jr., B. C.; Iacobucci, G. A. *J. Chromatog.* **1980**, *202*, 127-130.
5. Chamblee, T. S.; Clark, Jr., B. C.; Radford, T.; Iacobucci, G. A. *J. Chromatog.* **1985**, *330*, 141-151.
6. Clark, Jr., B. C.; Chamblee, T. S.; Iacobucci, G. A. *J. Agric. Food Chem.* **1987**, *35*, 514-518.
7. Shaw, P. E. *J. Agric. Food Chem.* **1979**, *27*, 246-257.
8. Chamblee, T. S.; Clark, Jr., B. C.; Brewster, G. B.; Radford, T.; Iacobucci, G. A. *J. Agric. Food Chem.* **1991**, *39*, 162-169.

9.　Clark, Jr., B. C.; Chamblee, T. S. In *Off-Flavors In Foods And Beverages*; Charalambous, G., Ed.; Elsevier: Amsterdam, 1992; pp 229-285.
10.　Rothe, M.; Thomas, B. *Z. Lebensm. Unters. Forsch.* **1963**, *202*, 302-310.
11.　Guadagni, D. G.; Buttery, R. G.; Harris, J. *J. Sci. Food Agric.* **1966**, *17*, 142-144.
12.　Teranishi, R.; Buttery, R. G.; Stern, D. J.; Takeoka, G. *Lebensm. Wiss. Technol.* **1991**, *24*, 1-5, and references therein.
13.　Sugisawa, H.; Takeda, M; Yang, R. H.; Takagi, N. *Nippon Shokuhin Kogyo Gakkaishi*, **1991**, *38(8)*, 668-674 and references therein.
14.　Casimir, D. J.; Whitfield, F. B. *Int. Fruchtsaft-Union. Wiss. Tech. Komm. Ber.* **1978**, *15*, 325-345.
15.　Drawert F.; Christoph, N. In *Analysis of Volatiles*; Schreier, P., Ed.; Walter de Gruyter & Co.: Berlin, 1984; pp 269-291.
16.　Christoph, N.; Drawert, F. In *Topics in Flavour Research*; Berger, R. G.; Nitz, S.; Schreier, P., Eds.; H. Eichhorn: Marzling-Hangenham, Germany 1985; pp 59-77.
17.　Ulrich, F.; Grosch, W. *Z. Lebensm. Unters. Forsch.* **1987**, *184*, 277-282.
18.　Schieberle, P.; Grosch, W. *J. Agric. Food Chem.* **1988**, *36*, 797-800.
19.　Schieberle, P.; Grosch, W. *Z. Lebensm. Unters. Forsch.* **1989**, *189*, 26-31 and references therein.
20.　Acree, T. E.; Barnard, J.; Cunningham, D. G. *Food Chem.* **1984**, *14*, 273-286.
21.　Cunningham, D. G.; Acree, T. E.; Barnard, J.; Butts, R. M.; Braell, P. A. *Food Chem.* **1986**, *19*, 137-147.
22.　Acree, T. E.; Cottrell, T. H. E. In *Alcoholic Beverages*; Birch, G. G.; Lindley, M. G., Eds.; Elsevier: New York, 1985; pp 145-159.
23.　Rothe, M. In *Proc. Int. Symp. Aroma Research*; Maarse, H.; Groenen, P.J., Eds.; Centre for Agricultural Publishing and Documentation (Pudoc): Wageningen, The Netherlands, 1975; pp 111-119.
24.　Meilgaard, M.C. In *Geruch und Geschmackstoffe, Int. Symp.*; Drawert, F., Ed.; Verlag Hans Carl: Nürnberg, 1975; pp 211-254.
25.　Thomas, A. F.; Bessiere, Y. *Nat. Prod. Rep.* **1989**, *6*, 291-309.
26.　Guadagni, D. G.; Buttery, R. G.; Okano, S.; Burr, H. K. *Nature* **1963**, *200*, 1288-1289.
27.　Baker, R. A. *Ann. N.Y. Acad. Sci.* **1964**, *116*, 495-503.
28.　Langler, J. E.; Day, E. A. *J. Dairy Sci.* **1964**, 1291-1296.
29.　Forss, D. A. In *Flavor Research, Recent Advances*; Teranishi, R.; Flath, R. A.; Sugisawa, H., Eds; Marcel Dekker: New York, 1981; pp 125-174.
30.　Laska, M.; Hudson, R. *Chem. Senses* **1991**, *16(6)*, 651-662 and references therein.
31.　von Sydow, E.; Andersson, J.; Anjou, K.; Karlsson, G. *Lebensm. Wiss. Technol.* **1970**, *3*, 11-17.
32.　*Volatile Compounds in Foods--Qualitative and Quantitative Data*, 6th ed.; Maarse, H.; Visscher, C. A.; Willemsens, L. C.; Boelens, M. H., Eds.; TNO-CIVO Food Analysis Institute: Zeist, The Netherlands, 1989, Vol. 1; pp 57-61.
33.　Ohloff, G. *Perfum. Flavor.* **1978**, *3(1)*, 10, 12, 14-22.
34.　Pelosi, P.; Pisanelli, A. M. *Chem. Senses* **1981**, *6(2)*, 87-93.

RECEIVED August 19, 1992

Chapter 9

Volatile Compounds from Japanese Marine Brown Algae

Tadahiko Kajiwara, Kazuya Kodama, Akikazu Hatanaka, and Kenji Matsui

Department of Agricultural Chemistry, Yamaguchi University,
Yamaguchi 1677–1, 753 Japan

Characteristic odoriferous compounds such as dictyopterenes and related hydrocarbons, sesquiterpenes and long chain aldehydes have been identified from wet and undecomposed brown algae along the coast of Japan as constituents of ocean smell and flavors reminiscent of algae. The physiological roles of volatile compounds *e.g.* male gamete-attracting activities of dictyopterenes, are described. Based on detection of possible biogenetic intermediates, the biogeneses of the long chain aldehydes and dictyopterenes from fatty acids are discussed.

Since ancient days, the aromas of marine algae have attracted the attention of man. However, the content of essential oils in the algae is much lower than that of terrestrial plants. The volatile compounds in dried algae were studied at an early development of analytical chemistry (*1*). Unfortunately, as characteristic compounds had not been identified, further development in this field was hindered. Odoriferous characteristic compounds of wet and undecomposed algae have been reported by Moore *et al.* in 1968 (*2*). Recently, the study of volatile compounds in wet and fresh marine algae has developed rapidly and its achievements have attracted the attention of chemists and biologists. Characteristic volatiles such as dictyopterenes and the related C_{11}-compounds (Figure 1), sesquiterpenes (Figure 2), and long chain fatty aldehydes have been identified as constituents of ocean smell and male gamete-attracting substances, and flavors reminiscent of algae (*3-9*). This paper will focus on our recent developments on identification and biogenesis of the characteristic volatile compounds in Japanese brown algae: Laminaliaceae, Alariaceae, Dictyotaceae, Scytosiphonaceae and Chordariaceae (*6-9*).

Flavors of Edible Algae

Marine algae are widely distributed in a world of waters. They are consumed in Japan, China, Korea, South Australia, New Zealand, Polynesia, and South America.

0097–6156/93/0525–0103$06.00/0

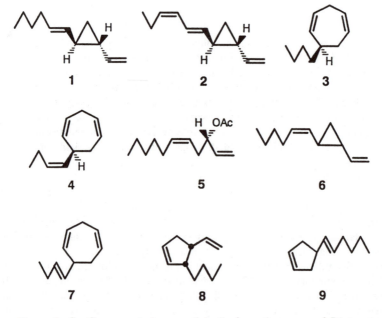

Figure 1. C_{11}-Compounds in essential oils from the genus of *Dictyopteris*.

Cadinane

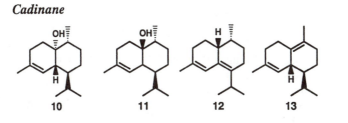

10   11   12   13

Tricyclic

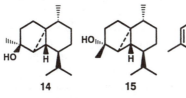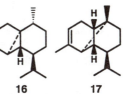

14   15   16   17

Elemane   *Germacrane*

18   19   20

Eudesmane   *Caryophyllane*

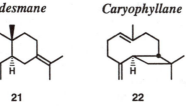

21   22

Figure 2. Sesquiterpenes in essential oils from the genus of *Dictyopteris*.

Many kinds of kelps, the great brown algae of northern waters, are cultivated on the coast of Japan for use mainly for food and occasionally for chemical algin.   A group of the kelps which belong to the genus *Laminaria, Kjellmaniella, Costaria, Ecklonia* and *Alaria,* and another group belonging to *Undaria* are generally respectively called "kombu" and "wakame" in Japanese.   These kelps have been eaten from antiquity and are highly favored for their tastes, aromas and textures.

Recently, essential oils of the wet and undecomposed edible kelps, *L. angustata, L. japonica, K. crassifolia, C. costata, A. crassifolia, E. cava* and *U. pinnatifida* along the Sea of Japan, were prepared by steam distillation. Volatile compounds in the essential oils were analyzed by gas chromatography (GC) and gas chromatography-mass spectrometry (GC-MS) equipped with fused silica glass capillary columns.   Fifty three compounds including alcohols, aldehydes, esters, ketones, hydrocarbons, and carboxylic acids were identified by comparison of Kovats indices and MS data with those of authentic compounds (Table I).   Volatile compounds of some air-dried kelps (*L. sp.*) which had been reported at an early stage of GC development (*10*), were not detected in the volatiles from the fresh and wet kelps, except for, 1-octen-3-ol.   Thus, the compounds in Table I were identified for the first time in the kelp oils with the exception of myristic and palmitic acids.   The sesquiterpene alcohol, cubenol (**10**), was detected in the volatile oils of all of the submitted kelps, whereas the stereoisomer, epicubenol (**11**) was detected only in *L. japonica* and *K. crassifolia* in small amounts.   In *U. pinnatifida* only two compounds, β-ionone and **10**, were detected.   The sesquiterpene alcohol (**10**), comprised *ca.* 90 % of the volatiles.   This is the first identification of **10** in kelps (*8*), and is also found in a Japanese brown alga, *Dictyopteris divaricata* (*11*).   The odor of the sesquiterpene (**10**) gave the flavorists the following images; kelp, hay, mint, and ocean.   Cubenol (**10**) had a medium odor intensity, and an odor threshold in the range of 100-250 ppm (in ethanol).   The odor profiles of kelp oils and descriptions of flavorists were affected by the compounds identified as follows; *L. angustata*, kelp with chicken fat note — (2*E*)-nonenal, (2*E*)-nonenol, and γ-nonalactone; *L. japonica*, standard kelp note; *K. crassifolia*, kelp with sweet and fat note — **10**, (2*E*)-nonenol; *C. costata*, kelp with sweet and citrus note — β-ionone, 1-octen-3-ol; *E. cava*, kelp with shell note — dimethyl-quinoline and others; *A.crassifolia*, kelp with aldehydic note — (2*E*,6*Z*)- nonadienal, (2*E*)-hexenal; *U. pinnatifida*, kelp with sweet note — **10**, β-ionone (*8*). From odor evaluation by sniffing tests, **10** was considered to have the characteristic flavor of kelp with a hay note, because the odor of **10** most resembled that of the submitted oils of kelps, and further, because no other compounds in Table I showed such kelp-like odor. (2*E*,6*Z*)-Nonadienal, (3*Z*,6*Z*)-nonadienal, (2*E*)-nonenal, and the corresponding alcohols, which are well known as flavor constituents of cucumber (*12,13*) and melons (*14,15*), were found to be the principle contributors of odors in some brown algae.   The C_9-aldehydes and alcohols particularly were at their highest concentration in *A. crassifolia*.   Some unsaturated fatty aldehydes and alcohols; *e.g.*

Table I. Volatile Compounds Identified in Essential Oils from *Laminariales*

Compounds	LA	LJ	KC	CC	EC	AC	UP
Alcohol							
(3Z)-Hexenol	+	-	-	-	-	+	-
(3E)-Hexenol	-	-	-	-	-	+	-
(2E)-Hexenol	-	+	-	-	-	+	-
Hexanol	+	+	-	+	-	+	-
1-Octen-3-ol	+	+	-	+	-	+	-
3-Octanol	+	-	-	-	-	-	-
(2E)-Octenol	+	+	+	+	-	+	-
(2E,6Z)-Nonadienol	+	-	+	-	-	++	-
(2E)-Nonenol	++	-	++	-	-	-	-
α-Terpineol	-	+	+	-	-	-	-
Decanol	+	-	-	-	-	-	-
(2E)-Decenol	-	+	-	-	-	-	-
Tridecanol	+	-	-	+	-	+	-
Epicubenol	-	+	+	-	-	-	-
Cubenol	++	++	++	++	++	+	++
α-Cadinol	+	-	++	+	-	-	-
Phytol	+	+	+	+	-	+	-
Aldehyde							
Hexanal	+	+	-	+	-	-	-
(2E)-Hexenal	+	+	-	+	-	+	-
Heptanal	+	-	-	+	-	-	-
(2E)-Heptenal	+	-	-	+	-	-	-
(2E,4E)-Heptadienal	+	+	-	+	-	+	-
(2E)-Octenal	+	+	-	+	-	+	-
(2E,4E)-Octadienal	-	+	-	+	-	-	-
(2E,4Z)-Nonadienal	+	+	+	+	-	++	-
(2E)-Nonenal	++	+	+	+	-	+	-
(3Z,6Z)-Nonadienal	-	-	-	-	-	++	-
(2E,4E)-Nonadienal	+	-	-	-	-	-	-
β-Cyclocitral	-	+	-	+	-	+	-
β-Homocyclocitral	-	+	-	-	-	+	-
(2E,4Z)-Decadienal	+	-	-	+	-	-	-
(2E,4E)-Decadienal	+	+	+	+	-	+	-
Dodecanal	+	-	-	-	-	-	-
Tridecanal	+	-	+	+	+	-	-
Pentadecanal	-	-	+	-	+	-	-
Ester							
γ-Nonalactone	+	-	-	-	-	-	-
Methyl dodecanoate	-	-	+	-	-	-	-
Dibutyl phtalate	+	+	++	+	-	+	-
Ketone							
1-Octen-3-one	+	-	-	+	-	-	-
β-Ionone	+	+	+	++	+	+	++
Hydrocarbon							
Styrene	-	-	+	-	-	-	-
Xylene	-	+	-	-	-	+	-
Butylbenzene	+	+	+	+	-	-	-
Naphthalene	+	-	+	-	-	-	-
γ-Elemene	-	-	+	-	-	-	-
Pentadecane	+	+	+	+	-	-	-
Acid							
Myristic acid	++	++	+	++	++	++	-
ω-Hexadecenoic acid	+	++	-	+	++	-	-
Palmitic acid	+	++	+	++	++	+	-
Oleic acid	+	-	-	+	-	-	-
Others							
Benzothiazole	+	-	+	-	+	-	-
Dimethylquinoline	-	-	-	-	+	-	-
Dihydroactinidiolide	-	-	-	+	-	+	-

LA: *Laminaria angusta*, LJ: *Laminaria japonica*, KC: *Kjellmaniella crassifolia*, CC: *Costaria costata*, EC: *Ecklonia cava*, AC: *Alaria crassifolia*, UP: *Undaria pinnatifida*.
-: not detected, +: detected, ++: major compound.

nonadienals, nonenals, and the corresponding alcohols are known to be green note compounds (16). The C_9-aldehydes and alcohols have been shown to contribute to a characterizing aroma from fish (17). (2E)-Nonenal and (2E,6Z)-nonadienal are formed via (3Z)-nonenal and (3Z,6Z)-nonadienal from 9-hydroperoxy-(10E, 12Z)-octadecadienoic acid and 9-hydroperoxy-(10E,12Z,15Z)-octadecatrienoic acid in higher plants, respectively (18,19). However, the biogenesis of the aldehydes in marine algae has not been studied. The nor-carotenoids such as β-cyclocitral, β-homocyclocitral, β-ionone, and dihydroactinidiolide, which had been reported as flavor components of an edible red alga, Porphyra tenera (20) seem to be important constituents of some brown algae such as C. costata and A. crassifolia.

Aromas of the Genus of Dictyopteris

In the earliest study of marine-derived sesquiterpenoids, Japanese researchers had steam-distilled the Japanese brown alga, D. divaricata, and obtained an oil with a "beach odor" (21). The oil has been reexamined and several sesquiterpenes have been identified ten years later (22). In striking contrast to the species of Dictyopteris, D. divaricata from Japan and D. zonarioides from Lower California (22,23), novel non-isoprenoid C_{11}-compounds (1-5) e.g. dictyopterenes have been identified in the essential oils of the Hawaiian species, D. plagiogramma and D. australis, instead of sesquiterpenes (e.g. 12 and 13) (24). This prominent difference in volatile compounds of the genus led us to explore both C_{11}-compounds (1-9) and sesquiterpenes (10-22) in the essential oils from this Japanese genus, D. prolifera, D. undulata and D. divaricata on the basis of GC and GC-MS analyses using a capillary column.

Compositions of C_{11}-Compounds and Sesquiterpenes. The two oils obtained from the steam distillate of fresh fronds of D. prolifera and D. divaricata displayed strikingly different chromatographic profiles. The C_{11}-hydrocarbons and sesquiterpenes identified are summarized in Table II. Dictyopterenes, zonarene, cubenol and epicubenol were identified by GC-MS, IR and NMR data. The other compounds were identified by comparison of Kovats indices and MS data with those of authentic specimens. With the essential oil of D. divaricata, a characteristic sesquiterpene alcohol, cubenol (10), comprised ca. 90% of the oil and no C_{11}-hydrocarbons were detected. In the essential oil of D. prolifera, cis-3-butyl-4-vinylcyclopentene (8), trans-1-((1Z)-hexenyl)-2- vinylcyclopropane (6), 4-((1E)-hexenyl)-cyclopentene (9) and 6-((1E)-butenyl)- cyclohepta-1,4-diene (7) together with the known C_{11}-compounds: dictyopterene A (trans-1-((1E)-hexenyl)-2-vinylcyclopropane 1), dictyopterene B (trans-1-((1E,3Z)-hexadienyl)-2-vinylcyclopropane 2), dictyopterene C' (6-butyl-1,4- cycloheptadiene 3), dictyopterene D' (6-((1Z)-butenyl)-1,4-cycloheptadiene 4), and dictyoprolene ((3S)-acetoxy-(1,5Z)-undecadiene 5) (25,26), were newly identified by comparison of retention indices of GC and mass spectra of GC-MS with those of authentic specimens synthesized through unequivocal routes. The three C_{11}-hydrocarbons (7-9) have been identified from the European member of this genus, D. membranacea

Table II. C₁₁-Compounds and Sesquiterpenes Identified in Essential Oils from the Genus of *Dictyopteris*

Compounds		*Dictyopteris*					
		DP	DU	DD	DZ	DL	D.sp.
Non-isoprenoid							
Dictyopterene A	**(1)**	++	+	-	-	++	+
Dictyopterene D'	**(4)**	++	+	-	-	+	+
Dictyopterene C'	**(3)**	+	+	-	-	+	+
Fatty acid derivatives		+	+	++	+	++	+
Isoprenoid							
Zonarene	**(12)**	-	++	-	++	+	++
Cubenol	**(10)**	+	+	++	+	+	+
Epicubenol	**(11)**	+	+	-	+	-	+
α-Cadinene	**(13)**	+	+	+	-	-	++
β-Elemene	**(18)**	+	+	+	+	-	+
δ-Elemene	**(19)**	+	+	-	++	-	+
Germacrene D	**(20)**	+	+	+	+	-	+
γ-Selinene	**(21)**	-	+	-	+	-	+
β-Caryophyllene	**(22)**	+	+	-	+	-	+
α-Cubebene	**(16)**	+	+	+	-	-	+
α-Copaene	**(17)**	-	++	+	-	-	+
Epicubebol	**(15)**	-	-	+	-	-	-

DP: *Dictyopteris prolifera*, DU: *D.undulata*, DD: *D.divaricata*, DZ: *D. zonarioides*, DL: *D. latiuscula*, D.sp.: an identified species of *Dictyopteris*.
-: not detected, +: detected, ++: major component.

(Stock.) Batt (27). However, *trans*-1-((1Z)-hexenyl)- 2-vinylcyclopropane (6) was identified for the first time as a natural product of *D. prolifera*. Particularly, the occurrence of the (Z)-isomer (6) of dictyopterene A (1) was also confirmed in the volatiles extracted by the closed-loop-stripping technique without heating. Dictyopterenes (2 and 3) were not resolved well on DB-1 column (0.25 mmϕ × 50 m), whereas 2 (*ca*. 80%), and 3 (*ca*. 20%) were isolated by a $AgNO_3$-impregnated silica gel column chromatography of the steam distillate from *D. prolifera*. Thus, the significant isomerization of dictyopterene B (2) to dictyopterene D' (4) did not occur during steam-distillation. However, $AgNO_3$ and/or steam-distillation could cause rearrangement and/or decomposition of some natural components (7).

Growth-locality Variation in Chemical Compositions. The volatile compositions in essential oils of *D. prolifera* and *D. divaricata*, which were collected from six locations in Japan (Muroran; the Pacific coast of Hokkaido in northern Japan, Osaka; inland-sea along the Pacific coast of Mid-Japan, Susa; the Japan coast of the northern Yamaguchi in southern Japan, Yoshimo; on the Japan coast of mid-Yamaguchi, Hikoshima; inland-sea along the Japan coast of southern Yamaguchi) were explored. With *D. prolifera*, the compositions of dictyopterene A (1) and dictyoprolene (5) were higher in the oils from the algae along the Japan coast, Susa and Yoshimo, than in those from Osaka and Hikoshima, whereas the compositions of dictyopterene B, C', and D' (2-4) were greater in the latter than in the former. The difference in compositions of the C_{11}-compounds might be due to the algal growing conditions, *e.g.* nutrient, mineral compositions, and temperature of sea water in the open-sea and in the inland-sea. The difference in compositions of 2-4 reflects particularly on the quality of "ocean smell of the essential oil" and the oil from the open-sea was found to possess more odor reminiscent of ocean than that from the inland-sea. On the other hand, a sesquiterpene alcohol, cubenol (10) was isolated from the steam distillate of fresh *D. divaricata* by a silica gel column chromatography. Instead of the C_{11}-compounds, over 90-95% of 10 were comprised in the oils from *D. divaricata* collected at two different places, northern Japan and southern Japan, along with small amounts of the structurally related sesquiterpenes, β-elemene (18) and germacrene D (20). Although only a few species have been studied, it seems that *Dictyopteris* species might be divided into two groups on the basis of their major constituents: (i) C_{11}-compounds, (ii) sesquiterpenes. On the other hand, the oil of *D. undulata* was found to consist of small amounts of the C_{11}-compounds (1-9) which were lacking in *D. divaricata*. Only traces of cubenol (10) was in *D. prolifera*. A characteristic sesquiterpene, zonarene (12) was found as the major constituent of *D. undulata* oil (7,28). Thus, another group of the genus was suggested in Japanese *Dictyopteris*. A survey of the literature showed that the essential oils of 4 from species of *Dictyopteris* have been chemically investigated. With *D. zonarioides* (*syn. D. undulata*) collected from lower California, the isolation of sesquiterpenes *e.g.* zonarene (12) has been reported, but the C_{11}-compounds have not been explored so far (23). With *D. membranaceae*, *D. plagiogramma*, and *D. australis*, the occurrence of the C_{11}-compounds was described, whereas sesquiterpenes in the oils have not been reported. Thus, the volatile compounds in the Californian *D. undulata* (*D. zonarioides*) were analyzed and the profile of the volatiles was compared with those of the Japanese species.   They showed very similar GC profiles and the characteristic sesquiterpene, zonarene (12) was identified as a major component in both the species, but the C_{11}-hydrocarbons were slight or non-existent in the Californian species.   The unidentified Japanese *Dictyopteris* in Table II, was suggested to be a close species to *D. undulata* from the GC profile and detection of small amounts of the C_{11}- compounds and 12.   Most recently, it has been found that epicubebol (15) comprises over 60-70% of the essential oil of *D. divaricata* in Osyoro along the Japan coast of Hokkaido, but less than 1% of the oil found in Muroran, which is on the Pacific coast.   Thus, further work on the C_{11}-compounds (1-9) and sesquiterpenes (10-22) is desirable, and seasonal variation of the volatile compounds in the genus *Dictyopteris* is being studied.

Volatile Compounds in Sexually Mature Thalli and Gametes

The odoriferous compounds in thalli and gametes of Japanese marine algae have been explored for elucidation of their physiological significance in marine ecological systems.   Characteristic volatile compounds were examined on the essential oils of sexually mature thalli of a northern Japanese brown alga, *Scytosiphon lomentaria*, and male and female gametes of *S. lomentaria*, *Colpomenia bullosa* and *Analipus japonicus*.   The essential oil from fresh sexually mature thalli of *S. lomentaria* was analyzed by GC and GC-MS.   The oil consisted mainly of sesquiterpenes (*ca.* 26%) such as cubenol (10), β-elemene (18) and β-ionone, and fatty acid metabolites (*ca.* 22%) such as hexanal, dodecanol and tridecanal (Table III).   As minor components, dictyopterene A (1), dictyopterene C' (3) and dictyopterene D' (4) were first identified by comparison of GC retention indices and mass spectra with those of synthetic specimens (*29,30,31*); 2 was isomerized to 4 during GC analysis under standard conditions (injection temp. 280°C, column temp. 70→220°C).   Thus, both 2 and 4 were identified in mature thalli of *S. lomentaria*.   To elucidate the origin of the dictyopterenes detected in the sexually matured thalli of *S. lomentaria*, female gametes of *S. lomentaria*, *C. bullosa* and *A. japonicus* prepared from the matured thalli are almost isomorphic.   The periods of swimming of the female gametes from *C. bullosa* and *A. japonicus* were shorter than those of *S. lomentaria*.   Female gametes settled down sooner than male gametes.   The difference between almost isomorphic female and male gametes was noticeable during their settling.   A pleasant fragrance emanated from the suspension of the female gametes of the two species when allowed to settle on a surface.   However, no smell could be detected from the swimming female gametes and both the settling and swimming male gametes in the three species.   Thus, the fragrance from settled female gametes was extracted using a closed-loop-stripping system (*32*).   After several hours of looping, the compounds adsorbed by an active carbon fiber were eluted with methylene chloride.   The eluates were then analyzed by GC, GC-MS and HPLC.   These showed the occurrence of dictyopterene B (2) in the secretions from the female gametes of *S. lomentaria*, *C. bullosa* and *A. japonicus*.   The secretions from both the female and male gametes of these species were also compared by HPLC analysis (Zorbax ODS, MeOH-H_2O-THF).   The highest peak was the characteristic peak of the secretion from female gametes, and other peaks were common to both female

Table III. Volatile Compounds in Sexually Mature Thalli, and
Female and Male Gametes of S. *lomentaria*

Compounds	Vegetative Thalli	Female Gametes	Male Gametes
Dictyopterene A	+	-	-
Dictyopterene D'	+	++	-
Dictyopterene C'	+	-	-
1,3,5,8-Undecatetraene	-	++	-
2,4,6,8-Undecatetraene	-	++	-
Tridecane	+	+	+
Tetradecane	+	+	+
Pentadecane	+	+	+
$C_{11}H_{20}O_2$ (M^+ 184)	-	++	-
(1,5Z)-Undecadien-3-yl-acetate	+	-	-
2-Pentenal	+	-	-
Hexanal	+	-	-
(2E)-Hexenal	+	-	-
(2E,4E)-Heptadienal	+	-	-
(2E,4E)-Nonadienal	+	-	-
Decadienal	+	-	-
Dodecanal	+	-	-
Tridecanal	++	-	-
Pentadecanal	++	-	-
2-Pentenol	+	-	-
1-Penten-3-ol	+	-	-
β-Cyclocitral	+	-	-
β-Ionone	++	+	+
δ-Cadinene	+	-	-
Cubenol	++	-	-

-:not detected, +: detected, ++: major component.

and male gametes (Figure 3). With female gametes of *C. bullosa*, the characteristic peak was isolated by HPLC (Zorbax ODS) and first identified as **2** by comparison of NMR (Figure 4), IR and MS data with those of the authentic specimen. The peak of the female secretions from *S. lomentaria* and *A. japonicus* was identified as **2** by comparison of GC-MS and HPLC data with those of a synthetic specimen. With one exception, a mixture of **2** (12%) and **4** (88%) was detected only in the female secretion of *A. japonicus*, accompanied by a slight amount of dictyopterene C' (**3**). But dictyopterenes A (**1**) and C' (**3**) detected in mature thalli of *S. lomentaria*, were not detected in either female or male gametes of *S. lomentaria* and *C. bullosa*.

Physiological Roles of Volatile Compounds. Müller *et al.* have reported that dictyopterene B (hormosirene **2**) is a sex-attractant of Australian *S. lomentaria* and *C. peregrina* (*33*). However, the configuration of **2** has not been reported so far. The absolute configuration of **2** secreted from female gametes of Japanese brown algae, *S. lomentaria* and *C. bullosa* was determined to be (1*R*,2*R*) by chiral HPLC analysis (Chiralcel-OB, cellulose tribenzoate coated on silica, Dicel Japan; eluent MeOH-H_2O), using synthetic (±)-dictyopterene B ((±)-**2**) and (-)-dictyopterene B ((-)-**2**) from *D. prolifera* (*29*). The optical purities (93-94% *e.e.*) of **2** in *S. lomentaria* and *C.bullosa* were much higher than those in *A. japonicus* (66% *e.e.*) (Table IV). The configuration of hormosirene (**2**) from *A. Japonicus* was determined to be (1*S*,2*S*). On the other hand, (1*R*,2*R*)-dictyopterene B ((-)-**2**) in vegetative thalli of *D. prolifera* and *D. undulata* were 90-93% *e.e.*. Müller *et al.* have obtained the similar results (*34,35*). These results in Table III suggest that species-specific enantiomeric mixtures of dictyopterene B (**2**) will be formed independently of sexuality or asexuality of thalli, and might be naturally secreted hormones of *S. lomentaria*, *C. bullosa* and *A. japonicus*. However, a mixture of **2** and **4** should be the sex-attractant of *A. japonicus*. Further quantitative tests of biological activity are necessary to reveal the ecological significance of the enantiomeric mixtures.

Biogenesis of Volatile Compounds

Few reports of biogenesis of marine algae-derived volatile compounds such as dimethyl sulfide, dictyopterenes, and long chain aldehydes are now available (*36-39*). The aldehydes (LCA) such as pentadecanal (PD), (8*Z*)-heptadecenal (HD), (8*Z*,11*Z*)-heptadecadienal (HDD) and (8*Z*,11*Z*,14*Z*)-heptadecatrienal (HDT) have been identified as characteristic flavors in the essential oils from fresh marine green algae belonging to Ulvaceae (*9*). Recently, these aldehydes (PD,HD,HDD,HDT) have been demonstrated to produce enzymatically from palmitic acid (PA), oleic acid (OA), linoleic acid (LA) and α-linolenic acid, respectively, in a green alga, *Ulva pertusa* and conchocelis-filaments of a red alga, *Porphyra tenera* (*37,38*). On the other hand, a model of biogenesis of odoriferous C_{11}-hydrocarbons such as algal pheromones, ectocarpene (dictyopterene D', **4**) and dictyotene (dictyopterene C', **3**) has been proposed using the terrestrial plant, *Senecio isatideus* (*39*).

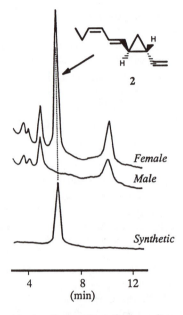

Figure 3. HPLC analysis of secretions from male and female gametes
of *C. bullosa.*

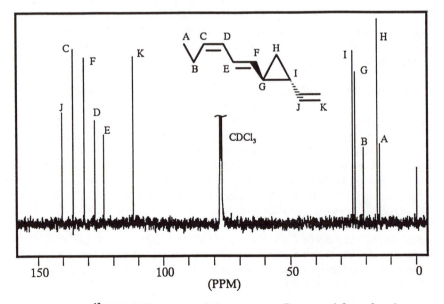

Figure 4. ^{13}C-NMR Spectrum of dictyopterene B secreted from female
gametes of *C. bullosa.*

Table IV. Enantiomeric Composition of Dictyopterene B in Female Gametes and Thalli of Brown Algae

Species	Locality	Major enantiomer	e.e.(%)
Scytosiphon lomentaria	Hikoshima (Japan)	(-)-1R,2R	94.0
Colpomenia bullosa	Murotsu (Japan)	(-)-1R,2R	92.0
Analipus japonicus	Muroran (Japan)	(+)-1S,2S	66.0
	Akkeshi (Japan)	(+)-1S,2S	90.0[2]
Dictyopteris prolifera[1]	Hikoshima (Japan)	(-)-1R,2R	90.0
Dictyopteris prolifera[1]	Susa (Japan)	(-)-1R,2R	92.0
Dictyopteris undulata[1]	Hikoshima (Japan)	(-)-1R,2R	92.0
Dictyopteris sp.[1]	Susa (Japan)	(-)-1R,2R	97.2
Dictyopteris membranaceae[1]	Villefranche (French Mediterr.coast)	(-)-1R,2R	71.2[2]
Durvillaea potatorum	Sorrento (Victoria Australia)	(-)-1R,2R	51.7[2]
Hormosira banksii	Flinders (Victoria Australia)	(-)-1R,2R	82.8[2]
Xiphophora chondrophylla	Flinders (Victoria Australia)	(-)-1R,2R	82.0[2]
Xiphophora gladiata	Hobart (Tasmania)	(-)-1R,2R	72.3[2]
Haplospora globosa	Halifax (Nova Scotia)	(+)-1S,2S	83.3[2]

1) Vegetative thalli.   2) Reproduced with permisson from reference *35*.

Biogenesis of Long Chain Aldehydes.   Long chain aldehydes such as pentadecanal, (8Z)-heptadecenal (HD), (8Z,11Z)-heptadecadienal (HDD) and (8Z,11Z,14Z)-heptadecatrienal (HDT), were first found in cucumber homogenates (40).   Recently, these aldehydes (HD,HDD,HDT) have been identified in the green marine algae Ulvaceae (9) as the major characteristic components of the essential oils. Galliard and Matthew (41) have reported the biogenesis of saturated C_{15}, C_{14}, C_{13} and C_{12}-aldehydes from palmitic acid in cucumber fruits.   Recently, the enzymatic formation of the saturated C_{15}-aldehyde and the unsaturated C_{17}-aldehydes (HD,HDD,HDT) in marine algae has been demonstrated.   On the other hand, a large mass of clean thalli culture regenerated from protoplasts of U. pertusa, was obtained by cultivation in Provasoli's enriched seawater containing antibiotics.   The essential oil of the thalli culture was obtained by a simultaneous distillation extraction (SDE) procedure (42).   The major characteristic compounds are long-chain aldehydes as in the oil from the field fronds: mainly n-pentadecanal, (7Z,10Z,13Z)-hexadecatrienal, HD, HDD, and HDT.   The long-chain aldehydes formed from unsaturated fatty acids by the enzymes of the thalli culture were the same as in the field fronds, i.e. HD from oleic acid, HDD from linoleic acid, HDT from α-linolenic acid.   Thus, it was first confirmed that the unsaturated C_{n-1}-aldehydes are produced from C_n-fatty acids by algal enzymes and not by the enzymes of attached bacteria or other epiphytes (43). Most recently, the aldehydes were found to be produced from fatty acids during the incubations of homogenates of brown algae, the genus of Dictyopteris.   HDD increased greatly, when LA was incubated with the enzyme solution of D. prolifera in phosphate buffer (pH 7.0).   Also HD and HDT were formed by incubations with OA and LNA, respectively.   With heat-treated suspensions, the increases were not observed.   Thus, long chain aldehyde-forming activity (LCAA) was first found in brown algae (44).   Most recently, a possible biogenetic intermediate of long chain aldehydes, 2-hydroperoxypalmitic acid was first detected during incubation of purified LCAA with palmitic acid (Kajiwara, T. unpublished).   Thus, a mechanism of long chain aldehydes formation via 2-hydroperoxycarboxylic acid was proposed as in Figure 5. This LCAA (α-oxidation) can peroxidatively decarboxylate a long fatty acid to produce a long chain aldehyde with one less carbon, which is oxidized to the corresponding acid by a NAD-dependent aldehyde dehydrogenase, and this acid again goes through the series of reactions until an inactive substrate is formed. LCAA (α-oxidation) is considered to be a key enzyme, which provides a precursor of dictyopterenes, as in Figure 7.

Biogenesis of Dictyopterenes.   Boland et al. have proposed the biogenesis of dictyopterenes via a possible precursor (3Z,6Z,9Z)-dodecatrienoic acid (Figure 6) using a terrestrial plant S. isatideus as a model system (4,39).   However, the C_{12}-trienoic acid has not been observed as yet. From quantitative HPLC analysis of dictyopterene B (2: λ_{max}=247 nm, ε=29,000), the secretion rate of 2 from female gametes of (60% settlement) was estimated to be 2.8×10^{-20} ±0 mol/cell/sec (n=3)

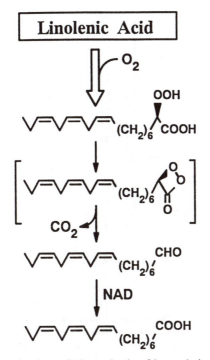

Figure 5. A mechanism of biosynthesis of long chain aldehydes *via* 2-hydroperoxides from fatty acids in brown algae.

in *S. lomentaria*, extraction efficiency was 49.5% (n=3) using synthetic (±)-2 (160 ng/ml) (*6*).   On the other hand, quantitative comparison of 2 in swimming female gametes with that of settled female gametes showed that the C_{11}-hydrocarbon (2) is biosynthesized in the process of the release of female gametes from conceptacles to the gamete settling.   Thus, the precursor of 2 in the male and female gametes of *A. japonicus* was explored in a form of 9-anthrylmethyl derivative of fatty acids by HPLC with a fluorescence detector (Ex. 365 nm; Em. 412 nm).   A possible biogenetic intermediate, (3Z,6Z,9Z)-dodecatrienoic acid was tentatively identified in female gametes, but not in male gametes by comparison of HPLC retention times and MS data [m/z(%) 348(15,$[M]^+$), 191(100)] with those of a synthetic specimen as shown in Figure 6 (Kajiwara, T. unpublished).   This detection strongly supports Boland's hypothesis (Figure 7) of biogenesis of dictyopterenes in marine brown algae.

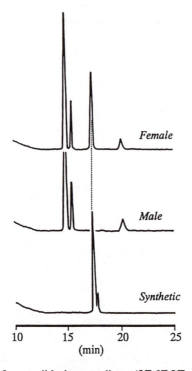

Figure 6. Detection of a possible intermediate, (3Z,6Z,9Z)-dodecatrienoic acid, in biogenesis of dictyopterene B secreted from female gametes of *A. japonicus*.
HPLC condition;   column, Zorbax C_8 (250 mm × 4.6 mmφ),
                   eluent, $CH_3CN:MeOH:H_2O$ (20:4:5),
                   flow rate, 1.1 ml/min.

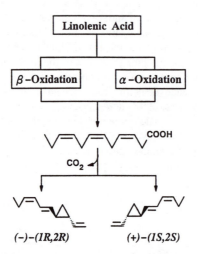

Figure 7. A biogenetic pathway of dictyopterene B from α-linolenic acid.

Acknowledgements. This work was supported in part by a Grant-in Aid (No. 01560153, 02556025 and 03806022) from the Ministry of Education, Science and Culture, Japan and a Grant from the Shorai Foundation for Science and Technology.

Literature cited

1. Katayama, T. In *Physiology and Biochemistry of Algae*; Lewin, R. A., Ed.; Academic Press: New York, 1962; pp 467-473.
2. Moore, R. E.; Pettus, J. A.Jr. *Tetrahedron Lett.* **1968**, 4789-4790.
3. Moore, R. E. *Account of Chem. Res.* **1977**, *10*, 40-47.
4. Boland, W. In *Bioflavour '87*; Schreier, P. Ed.; De Gruyter: Berlin, 1988; pp 199-220.
5. Müller, D. G.; Schmid, C. E. *Biol. Chem. Hoppe Seyler* **1988**, *369*, 647-653.
6. Kajiwara, T.; Hatanaka, A.; Kodama, K.; Ochi, S. *Phytochemistry* **1991**, *30*, 1805-1807.
7. Kajiwara, T.; Hatanaka, A.; Tanaka, Y.; Kawai, T.; Ishihara, M.; Tsuneya, T.; Fujimura, T. *Phytochemistry* **1989**, *28*, 636-638.
8. Kajiwara, T.; Hatanaka, A.; Kawai, T.; Ishihara, M.; Tsuneya, T. *J. Food Sci.* **1988**, *53*, 960-962.
9. Kajiwara, T.; Hatanaka, A.; Kawai, T.; Ishihara, M.; Tsuneya, T. *Bull. Japan. Soc. Sci. Fish.* **1987**, *53*, 1901.
10. Katayama, T. *Bull. Japan. Soc. Sci. Fish.* **1961**, *27*, 75-84.
11. Suzuki, M.; Kiwata, N.; Kurosawa, E. *Bull. Chem. Soc. Jpn.* **1981**, *54*, 2366-2368.
12. Fross, D. A.; Dunstone, E. A.; Ramshaw, E. H.; Stark, W. *J. Food Sci.* **1962**, *27*, 90-93.
13. Kemp, T. R.; Knavel, D. E.; Stoltz, L. P. *J. Agric. Food Chem.* **1974**, *22*, 717-718.
14. Kemp, T. R. *Phytochemistry* **1975**, *14*, 2637-2638.
15. Yajima, I.; Sakakibara, H.; Ide, J.; Yanai, T.; Hayashi, K. *Agric. Biol. Chem.* **1985**, *49*, 3145-3150.
16. Morris, A. F. *Perfumer and Flavorist* **1981**, *6*, 1-7.
17. Josephson, D. B.; Lindsay R. C. In *Biogeneration of Aromas*; Parliment, T. H.; Croteau, R. Eds.; *ACS Symp. Ser. 317*; American Chemical Society: Washington, DC, 1986; pp201-219.
18. Fleming, H. P.; Cobb, W. Y.; Etchells, J. L.; Bell, T. A. *J. Food Sci.* **1968**, *33*, 572-576.
19. Hatanaka, A.; Kajiwara, T.; Harada, T. *Phytochemistry* **1975**, *14*, 2589-2592.
20. Flament, I.; Ohloff, G. In *Progress in Flavour Research*; Adda, J. Ed.; Elsevier Science Publishers B. V.: Amsterdam, 1984; pp281-300.
21. Takaoka, M.; Ando, Y. *J. Chem. Soc. Jpn.* **1951**, *72*, 999-1003.
22. Irie, T.; Yamamoto, K.; Masamune, T. *Bull. Chem. Soc. Jpn.* **1964**, *37*, 1053-1055.
23. Fenical, W.; Sims, J. J. *Phytochemistry* **1972**, *11*, 1161-1163.
24. Moore, R. E.; Pettus, J. A. Jr.; Mistysyn, J. *J. Org. Chem.* **1974**, *39*, 2201-2207.

25. Kajiwara, T.; Kodama, K.; Hatanaka, A. *Bull. Japan. Soc. Sci. Fish.* **1980**, *46*, 771-775.
26. Yamada, K.; Tan, H.; Tatematsu, H.; Ojika, M. *Tetrahedron* **1986**, *42*, 3775-3780.
27. Boland, W.; Müller, D. G. *Tetrahedron Lett.* **1987**, *28*, 307-310.
28. Kajiwara, T.; Kodama. K.; Hatanaka, A. *Bull. Japan. Soc. Sci. Fish.* **1982**, *48*, 211-214.
29. Kajiwara, T.; Nakatomi, T.; Sasaki, Y.; Hatanaka, A. *Agric. Biol. Chem.* **1980**, *44*, 2099-2104.
30. Kajiwara, T.; Sasaki, Y.; Kimura, F.; Hatanaka, A. *Agric. Biol. Chem.* **1981**, *45*, 1461-1466.
31. Schotten, T.; Boland, W.; Jaenicke, L. *Helv. Chim. Acta,* **1985**, *68*, 1186-1192.
32. Grob, K.; Zürcher, F. *J. Chromatogr.* **1976**, *117*, 285-294.
33. Müller, D. G.; Clayton, M. N.; Gassmann, G.; Boland, W.; Marner, F. -J.; Schotten, T.; Jaenicke, L. *Naturwissenschaften* **1985**, *72*, 97-99.
34. Müller, D. G.; Kawai, H.; Stache, B.; Folster, E.; Boland, W. *Experientia* **1990**, *46*, 534-536.
35. Boland, W.; Flegel, U.; Jordt, G.; Müller, D. G. *Naturwissenschaften* **1987**, *74*, 448-449.
36. Cantoni, G. L.; Anderson, D. G. *J. Biol. Chem.* **1956**, *222*, 171-177.
37. Kajiwara, T.; Yoshikawa, H.; Saruwatari, T.; Hatanaka, A.; Kawai, T.; Ishihara, M.; Tsuneya, T. *Phytochemistry* **1988**, *27*, 1643-1645.
38. Kajiwara, T.; Kashibe, M.; Matsui, K.; Hatanaka, A. *Phytochemistry* **1990**, *29*, 2193-2195.
39. Neumann, C.; Boland, W. *Helv. Chim. Acta* **1990**, *73*, 754-761.
40. Kemp, T. R. *J. Am. Oil Chem. Soc.* **1975**, *52*, 300-302.
41. Galliard, T.; Matthew, J. A. *Biochem. Biophs. Acta* **1976**, *424*, 26-35.
42. Schultz, T. H.; Flath, R. A.; Mon, T. R.; Eggling, S. B.; Teranishi, R. *J. Agric. Food Chem.* **1977**, *25*, 446-449.
43. Fujimura, T.; Kawai, T.; Shiga, M.; Kajiwara, T. *Phytochemistry* **1990**, *29*, 745-747.
44. Kajiwara, T.; Tomoi, T.; Hatanaka, A.; Matsui, K. *Phytochemistry* in press.

RECEIVED August 19, 1992

Chapter 10

Aroma Profiles of Peel Oils of Acid Citrus

H. Tamura, R.-H. Yang, and H. Sugisawa

Department of Bioresource Science, Kagawa University, Miki-cho, Kagawa, 761–07 Japan

The aroma quality of six acid citrus fruits was characterized by instrumental and objective methods such as the GC-sniff test with flavor wheel and logarithmic values of odor units. With the GC-sniff test, carvone, carveol, perillaldeyde, linalool and citronellal were found to constitute the characteristic aroma of sudachi oil. Logarithmic odor unit values of each component of lemon, lime, sudachi, kabusu, yuzu, and kabosu were found to be effective for more objective evaluation of aroma quality. Monoterpene aldehydes, alcohols and their esters, for example, geranial, geraniol, and neral contributed significantly to the characteristic aroma of lemon and lime. Citronellal, citronellol, and carvone contributed to sudachi; and citronellol, 2-decenal, geranyl acetate to kabosu. Geraniol, linalyl acetate, geranyl acetate and linalool contributed to kabusu. Unsaturated aliphatic aldehydes especially contributed to yuzu. Multivariate analysis was applied to aroma evaluation of many samples.

There are several hundred species in the Citrus family. The sour and sweet taste offers a refreshing, pleasant and special sensation. Lemon and lime are representative citrus fruits all over the world and have a particular sour taste. In this article, the name "acid citrus" refers to a group of citrus fruits having sour juice. It is defined by the sour taste of these citrus fruits. Citrus flavor plays an important role in the promotion of our appetite, and acid citruses have been used as substitutes for vinegar in many Japanese foods. Yuzu, kabosu, sudachi and kabusu, as well as lemon and lime, are very common as acid citrus in Japan. These citrus fruits are present in many Oriental flavors. To ascertain the important compounds in the fruit aroma, identification of the volatile components, quantitative analysis and sensory evaluation of citrus oils have been performed. As new analytical techniques have been developed, trace constituents of the fruits were

0097–6156/93/0525–0121$06.00/0

identified. However, identification and quantitative analysis of all constituents in the fruits are not always essential for evaluation of flavor quality nor for the reproduction of the same flavor. A character impact compound, some character components or some composition of volatile compounds producing the unique aroma are significant. Recently a computer-aided analytical system for identification using retention index and similarity index of Mass spectral pattern has been helpful and indispensable.

To determine the significant compounds in each acid citrus fruit, the following two methods were examined and compared: 1. GC-sniff test using a flavor wheel, in which intensity and quality of each flavor component can be expressed easier by a simple word, and 2. multivariate statistical analysis using logarithmic values of odor units of each component identified.

The following citrus fruits grown on the island of Shikoku in Japan were used for this experiment: lemon (*Citrus limone*), lime (*Citrus aurantifolia*), sudachi (*Citrus sudachi* Hort. ex Shirai), kabosu (*Citrus sphaerocarpa* Tanaka), kabusu (*Citrus aurantium* f. kabusu), and yuzu (*Citrus junos* Tanaka).

Isolation and identification of volatile compounds in citrus fruits

Isolation of volatile compounds in 6 acid citrus fruits is the most critical step in the evaluation of aroma quality. Cold pressed oils from Valencia orange (*1*), grapefruit (*2*) and lime oil (*3*) were reported to have a quality very much like that obtained under mild conditions. However, the oils also included considerably higher amounts of non-volatile components. It is well-known that the characteristic aroma components in yuzu (*Citrus juno*) are thermally very unstable. The aroma qualities of yuzu oil extracted by three isolation methods - simultaneous distillation extraction (SDE) at atmospheric pressure, steam distillation under reduced pressure, and solvent extraction - were compared with those of fresh peels by paired comparison test. A panel consisting of six members carried out a series of sensory evaluations by duo-trio and paired comparison tests. This panel judged that solvent extraction produced material having an aroma similar to that of fresh yuzu peels. Volatiles from the other 4 acid citrus peels (400g) with pentane and dichloromethane (7/3, v/v) were also extracted by solvent extraction and SDE methods for 2 hours, respectively, and the concentrates were shown to retain the original peel aroma. The yield of each oil by solvent extraction was 0. 26% for lemon, 0.31% for lime, 0.10% for sudachi, 0.25% for yuzu, and 0.38% for kabosu. Volatiles of kabusu were extracted only by the SDE method (0.61% yield). Each oil was separated into hydrocarbon and oxygenated compounds fractions with a silica gel column and was analyzed by gas chromatography (GC) equipped with a flame ionization detector and gas chromatography-mass spectrometry (GC/MS). The oxygenated compounds had an aroma similar to those of the peels. Concentration of the volatiles in each oil was calculated from GC peak areas. Ethyl decanoate was added to each extract as an internal standard. The final concentration of ethyl decanoate was adjusted to 15ppm (w/w). For the purpose of identifying the volatile compounds, MS spectral patterns and retention indices of compounds identified in 6 acid citrus fruits were matched with those in the MS data library (containing about 35000 compounds) supplied by

NBS/NIH/EPA and those in our GC retention data library (containing 3900 compounds). The volatiles identified were: 72 compounds in lemon, 76 in lime, 69 in sudachi, 71 in yuzu, 69 in kabusu, and 77 in kabosu. Oxygenated compounds identified in each citrus peel oil are shown in Table I. Terpenoid compounds composed more than 76% of the oxygenated fractions of lemon, lime, yuzu, kabusu and sudachi peel oils. Lemon and lime peel oils contained mainly terpene aldehydes (>42%). On the other hand, yuzu, kabusu and sudachi contained terpene alcohols (>46%). About 65% of kabosu peel oil were aliphatic aldehydes.

Similarity of aroma quality was olfactorily evaluated by a panel of 6 members. As shown in Table II, lemon and lime formed one cluster and also showed most similar relationship followed by sudachi, kabusu and yuzu which had 41% similarity.

GC-sniff test using a flavor wheel

In 1976 a system was designed by Acree et al. (4) for sniffing effluents from gas chromatograph effluents. This method was developed as an apparatus for the measurement of human response to aroma and for finding character impact compounds. Aroma Extraction Dilution Analysis (5) and Charm Analysis (6) can be used to determine dilution values of each effluent. Nitz (7) also developed multidimensional gas chromatography in combination with sniffing-MS monitoring system. In addition, the GC sniff test with a flavor wheel was used for the characterization of important aromas in navel oranges (8). The advantage of the GC-sniff test, with the development of columns which can separate enantiomers, is that natural aroma compounds can be evaluated to determine the sensory properties of each enantiomer. In many cases, the important aroma of a terpenoid is a contribution by one enantiomer.

A flavor wheel, Figure 1, was used to record odor quality and intensity of each aroma constituent in acid citrus volatiles as each material was sniffed and sensory properties described as it was emerged from a GC sniff port. Prospective panel members were selected on the basis of their ability to distinguish standard aromas with a T & T olfactometer, available from Dai-ichi Yakuhin Sangyo Co., Ltd., Tokyo, Japan. Five standard aromas are tested at 8 different concentrations. The five standard aromas are: floral-like (β-phenylethyl alcohol), burnt-like (methyl cyclopentenolone), roten-like (iso-valeric acid), fruity-like (γ-undecalactone), and fecal-like (skatole).

Each individual was then trained for three months to become familiar with the compounds involved and the relationship of the sensory properties and the odor descriptions of these compounds on the odor wheel. For descriptive terms, in aroma profiles, four main terms were selected: floral-like, fruity-like, woody-like and herbaceous-like. Thirty-two terms were used to form the wheel. An odor was ranked on a 1 to 5 point scale according to its intensity: 1 = very weak aroma; 3 = medium aroma; 5 = extremely strong aroma. Two and four points were given when the intensity was judged between 1 and 3, or 3 and 5. Each aroma constituent eluted was sniffed at the sniff port under humid atmosphere.

Descriptive terms and the intensity of volatile components in oxygenated fraction of sudachi peel oil (9) are shown in Table III. Carvone, carveol,

Table I. Concentration and Odor Threshold of the Oxygenated Compounds in Peel Oils from Six Acid Citrus

No	Compound	Threshold (ppm)	Concentration (ppm)						Retention Index	
			Lemon	Lime	Sudachi	Yuzu	Kabosu	Kabusu	Found	Authentic Sample
1	hexanal	0.064			0.216	0.031	0.064	0.035	776	775
2	(E)-2-hexenal	0.48	0.337		0.285	0.327	0.383	0.248	827	827
3	(Z)-3-hexenol	3.625		2.167	0.827		0.758		834	836
4	(E)-2-hexenol	4.25				0.038	0.204		846	845
5	hexanol	3.125					0.117		849	848
6	heptanal	0.06							880	882
7	benzaldehyde	1.1						0.035	946	947
8	heptanol	0.425						0.035	949	950
9	6-methyl-5-hepten-2-one	1							962	965
10	butyl butyrate	0.55							975	977
11	octanal	0.082	1.217	0.463	6.204	0.506	5.290	5.103	980	981
12	(Z)-3-hexenyl acetate	0.32				0.067			983	986
13	hexyl acetate	0.48	0.018			0.122		0.070	991	993
14	benzyl alcohol	100							1007	1007
15	1,8-cineole	0.023	1.927	14.431	0.410	0.215		0.213	1024	1030
16	2,6-dimethyl-5-heptenal	0.016							1036	1039
17	octanol	0.875	0.621	0.458	0.307	0.105	1.231	10.278	1052	1053
18	trans-sabinene hydrate	55	6.982	15.958	4.213	1.489	0.283	7.123	1059	1059
19	cis-linalool oxide	0.32			0.374				1063	1068
20	fenchone	1.2	0.026						1072	1072
21	trans-linalool oxide	0.32			0.159	0.351	5.793	3.721	1076	1076
22	nonanal	0.1	3.903	34.065	3.261	0.641		1.736	1085	1084
23	linalool	0.028	13.808	24.520	71.372	119.611	3.763	94.448	1088	1088
24	heptyl acetate	0.32	0.048		4.352	0.243		0.354	1090	1094
25	trans-p-mentha-2,8-dien-1-ol	u			0.703	0.057		0.142	1110	1111
26	trans-p-menth-2-en-1-ol	u							1112	1111
27	cis-limonene oxide	0.25			1.540	0.088	0.380	0.354	1123	1122
28	cis-p-mentha-2,8-dien-1-ol	u							1124	1127
29	trans-limonene oxide	0.25			1.299	0.194	0.248		1125	1125
30	camphor	4.6	0.954	1.198				0.142	1125	1125
31	cis-p-menth-2-en-1-ol	u			0.450	0.330		0.390	1130	1128

No.	Compound								RI	RI
32	citronellal	0.046	4.597	8.747	5.812		0.244	0.248	1132	1135
33	(E)-2-nonenol	0.13	1.611	1.890	0.145		0.188		1153	1152
34	borneol	0.14	0.770						1153	1155
35	nonanol	1			1.004	0.085		2.410	1156	1156
36	4-isopropyl-2-cyclohexen-1-one	u							1165	1165
37	(Z)-3-hexenyl butyrate	0.32	2.554	6.807	1.679	2.016	0.429	1.950	1166	1167
38	terpinen-4-ol	6.4	27.583	71.880	29.230	10.287	1.875	31.223	1167	1169
39	α-terpineol	15	1.685	7.515	4.599	2.203	27.982	9.108	1178	1177
40	decanal	0.07	0.145	0.207		0.041	0.410	8.718	1184	1185
41	octyl acetate	0.45				0.138		0.815	1190	1193
42	cis-carveol	4	4.747	6.800	2.501		0.186		1198	1198
43	grandisol	u					0.714		1204	1200
44	citronellol	0.062			5.535	0.038			1209	1209
45	trans-carveol	4							1211	1211
46	nerol	0.68	23.975	31.993	5.300	0.277	0.744	8.541	1212	1211
47	neral	0.1	116.840	184.512	0.721	0.233	0.343	0.248	1218	1219
48	carvone	0.067		0.477			0.310	3.579	1220	1224
49	piperitone	u	0.332	42.280			0.353		1235	1236
50	geraniol	0.01	26.538	290.535		0.224	0.295	17.578	1236	1236
51	(E)-2-decenal	0.017							1240	1242
52	linalyl acetate	1	193.071					57.377	1242	1247
53	geranial	0.1			1.190	0.140	0.557	3.473	1246	1247
54	perillaldehyde	0.062	3.836	4.644	1.531	0.224	0.837	5.635	1253	1252
55	decanol	0.775	0.035	1.101			0.348		1262	1259
56	thymol	0.79	0.323	0.143	0.672	8.819			1268	1267
57	bornyl acetate	1.38	0.323	0.755				1.240	1274	1275
58	perill alcohol	7		0.569					1275	1276
59	carvacrol	2.29	0.144	0.221	0.317	0.202	0.422		1277	1278
60	geranyl formate	0.2	0.403	0.009						1279
61	undecanal	0.04	1.344	1.208	0.284	0.516	2.821	0.957	1287	1287
62	(E,E)-2,4-decadienal	0.01		0.135		0.228	0.123	0.142	1289	1292
63	nonyl acetate	0.6	0.161					1.453	1290	1292
64	methyl geranate	u	0.371						1299	1302
65	cis-carvyl acetate	u	0.057	0.243		0.046	0.141		1311	1315

Continued on next page

Table I. Continued

No	Compound	Threshold (ppm)	Concentration (ppm)						Retention Index	
			Lemon	Lime	Sudachi	Yuzu	Kabosu	Kabusu	Found	Authentic Sample
66	eugenol	0.1					0.066		1333	1330
67	terpinyl acetate	2.5	1.286	0.837				1.949	1333	1333
68	citronellyl acetate	1	38.150	1.759	0.082		0.233		1333	1336
69	neryl acetate	2	21.817	174.504			1.833	8.080	1341	1344
70	geranyl acetate	0.15		82.633			0.856	19.193	1359	1362
71	methyl eugenol	0.775					0.088		1371	1372
72	dodecanal	0.055	1.035	4.180	0.650	0.591	6.168	3.296	1387	1388
73	decyl acetate	0.225	0.049	0.438			0.265	3.509	1393	1394
74	geranylacetone	6.4					0.033		1425	1422
75	(E)-2-dodecenal	0.0014					0.128	1.559	1445	1444
76	β-ionole	0.006	0.258	0.663					1470	1464
77	tirdecanal	0.07	0.080				0.265		1488	1490
78	undecyl acetate	0.25			0.237	0.074		0.071	1493	1495
79	elemol	0.1	0.107	0.863	0.529	0.502	0.112		1543	1540
80	nerolidol	2.25					0.465		1548	1551
81	caryophylene oxide	5.5					0.144	6.131	1582	1576
82	tetradecanal	0.067	0.091	2.506	0.168		0.148		1592	1591
83	eudesmol	u			1.410	0.354			1644	1644
84	7-methoxy-2H-1-benzopyran-2-one	u		31.210					1671	
85	β-sinensal	0.082							1672	1671
86	pentadecanal	u	0.099	1.261		0.206		0.248	1690	1692
87	farnesol	u							1698	1698
88	nootkatone	0.28	0.486	3.214				1.630	1780	1775
89	hexadecanal	u	0.224	12.304				0.602	1794	1794
90	heptadecanal	u	0.203	1.865					1894	1895
91	5,7-dimethoxy-2H-1-benzopyran-2-one	u	1.017	13.527			0.493		1916	
92	9-methoxy-7H-furo[3,2-g]-[1]-benzopyran-7-one	u		5.199					1990	

Table II. Similarity (%) of Odor Quality in the Peel Oils from Five Acid Citrus Fruits judged by sensory evaluation

Citrus	Lemon	Lime	Sudachi	Yuzu
Kabosu	20	20	45	41
Yuzu	23	21	43	
Sudachi	25	29		
Lime	73			

Reproduced with permission from reference 15.

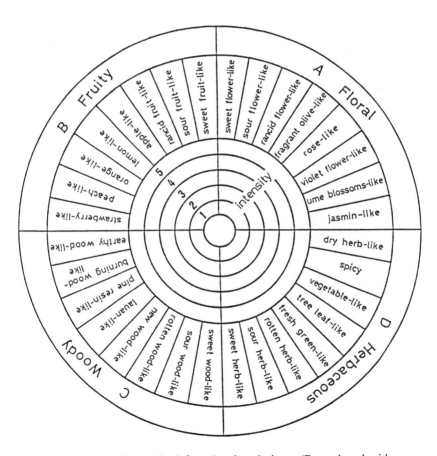

Figure 1. Flavor wheel for odor description. (Reproduced with permission from reference 8).

Table III. GC-Sniff Analysis of the Oxygenated Fraction in Sudachi Peel Oil

Peak No.	Odor description	Intensity*	Compound
1	Dry herb-like	1	
2	Rotten herb-like	1	
3	Fresh green-like	3	hexanal
4	Fresh green-like	1	
5	Rotten herb-like	1	
6	Sweet flower-like	1	
7	Sweet flower-like	2	
8	Orange-like	1	
9	Sweet flower-like	4	octanal
10	Burnig wood-like	1	
11	Spicy	2	
12	Sweet fruit-like	3	cis-linalool oxide
13	Sweet wood-like	1	
14	Sweet flower-like	5	nonanal, linalool
15	Spicy	2	
16	Fresh green-like	2	
17	Fresh green-like	3	cis-limonene oxide
18	Sour fruit-like	1	
19	Sour flower-like	3	citronellal
20	Spicy	2	
21	Earthy wood-like	2	
22	Earthy wood-like	3	
23	Orange-like	4	decanal
24	Rotten herb-like	3	
25	Sweet flower-like	3	citronellol
26	Pepermint-like	3	carvone
27	Pepermint-like	3	piperitone
28	Lemon-like	2	
29	Shiso-like	3	perillaldehyde
30	Fresh green-like	2	
31	Tree leaf-like	1	
32	Sweet flower-like	2	
33	Sour fruit-like	2	
34	Rotten herb-like	2	

*Intensity, 1=very weak ~ 5=very strong.

perillaldeyde, linalool and citronellal had aroma intensities greater than 3, and therefore were considered to contribute significantly to the sensory character of sudachi oil. A synthetic mixture of these compounds had a sudachi-like aroma but did not have the complete characteristic aroma quality of sudachi peel oil. Other compounds are still needed to approximate the sudachi oil.

Logarithmic values of odor units of each component of acid citrus fruits.

To pick up the significant conponents contributing to each characteristic aroma in 6 acid citrus fruits, the relative contribution of individual components to the overall odor intensity of the total volatiles was estimated by another method. The "odor-unit", according to the procedure of Guadagni et al. (*10*) and Rothe and Thomas (*11*) was used. The odor-unit (Uo) of the individual components is expressed as the following equation (1), as defined by Guadagni et. al. (*10*):

$$\text{Odor unit (Uo)} \; = \; \frac{\text{concentration of the volatiles in food (ppm)}}{\text{concentration at threshold (ppm)}} \tag{1}$$

The odor-threshold of each component was determined by Shibamoto's method (*12*). Two cups of five contained aqueous solutions of volatile compounds; the other three contained pure water. Thus, the six members of the sensory panel should be able to determine which two cups contain the volatile compounds. Each aqueous solution of volatiles was diluted by a factor of two until the solution was judged as odorless. The concentration at which panel members could no longer detect the compounds is defined as the odor threshold. The odor threshold of 73 compounds identified in acid citrus fruits was measured by 6 panel members. The value of each threshold was measured twice in 1988 and 1990, independently. Although there were some changes in panel members in two years, the threshold values remained relatively constant. The odor-thresholds and logarithmimc values of odor units of the components of the acid citrus fruits are listed in Table I and Table IV. Compounds having greater than 1 odor unit value are assumed to contribute to the characteristic aroma of the acid citrus. Compounds having less than 1 odor unit value are assumed not to contribute to the characteristic aroma. The odor unit value is to be used only as a first approximation or indication. In a strict sense, compounds with low odor unit values should be considered in adding to the delicate nuances of a given characteristic aroma.   However, volatile components which have large odor unit values contribute signnificantly to the total aroma quality of each fruit. Recently Buttery et al. (*13*) showed the importance of logarithmic values of odor units (log [Uo]) in the evaluation of tomato flavor. Sugisawa et al. (*14*) also applied this method to the evaluation of the aroma in navel oranges. In both cases, compounds with large positive values of log [Uo] proved to be important in contributing to the characteristic sensory properties.

Figure 2 shows the logarithmic values of concentrations and odor units of the oxygenated compounds in sudachi peel oil. The compounds in sudachi peel oil which were chosen as the important flavor components by the GC-sniff test with the flavor wheel were the same compounds which were judged to be important bv

Table IV. Log [Uo] of the Oxygenated Compounds in the Peel Oils from Six Acid Citrus

No.	Compound	Log[Uo]					
		Lemon	Lime	Sudachi	Yuzu	Kabosu	Kabusu
1	hexanal			0.5			
11	octanal	1.2	0.8	1.9	0.8	1.8	1.8
15	1,8-cineol	1.9	2.8	1.3	1.0		0.5
17	octanol					0.1	1.0
19	cis-linalool oxide			0.1			
22	nonanal	1.6	2.5	1.5	0.8	1.8	1.2
23	linalool	2.7	2.9	3.4	3.6	2.1	3.5
27	cis-limonene oxide			0.8		0.2	
29	trans-limonene oxide			0.7			
32	citronellal	2.0	2.3	2.1		0.7	0.4
33	(E)-2-nonenol			0		0.2	
34	borneol	1.1	1.1				
35	nonanol						0.4
39	a-terpineol	0.3	0.7	0.3			0.5
40	decanal	1.4	2.0	1.8	1.5	2.6	1.3
41	octyl acetate						0.3
44	citronellol	1.9	2.0	2.0		1.1	
46	nerol	1.5	1.7				0.6
47	neral	3.1	3.3			0.5	
48	carvone			1.9	0.6	0.7	0.6
50	geraniol	3.4	3.6				2.2
51	(E)-2-decenal				1.1	1.2	
52	linalyl acetate						1.8
53	geranial	3.3	3.5	1.1		0.7	
54	perillaldehyde	1.8	1.9	1.4	0.4	1.1	1.5
55	decanol		0.2				0.3
56	thymol				1.0		
60	geranyl formate	0.3					
61	undecanal	1.5	1.5	0.9	1.1	1.8	1.0
62	(E,E)-2,4-decadienal		1.1		1.4		1.2
63	nonyl acetate						0.4
68	citronellyl acetate	0.1	0.2				
69	neryl acetate	1.3	1.9				
70	geranyl acetate	2.2	2.7			0.8	1.6
72	dodecanal	1.3	1.9	1.1	1.0	2.0	1.8
73	decyl acetate		0.3			0.1	0.5
75	(E)-2-dodecenal				2.0		
77	tridecanal	0.6	1.0			0.6	
79	elemol			0.4			
82	tetradecanal	0.1	1.6	0.4		0.3	
88	nootkatone	0.2	1.1				0.8

the logarithmic values of odor units. Log [Uo] values of the 6 acid citrus fruits, yuzu, sudachi, lemon, lime, kabusu, and kabosu, are shown in Table IV. The compounds in Table IV having more than zero odor unit value in the all acid citrus would contribute to the citrus-like aroma. Seven compounds - octanal, nonanal, decanal, undecanal, dodecanal, linalool, perillaldehyde - were found to be the basis of general citrus aroma. The other compounds of log [Uo] > 0 must impart the individual character in each given aroma.

In these studies, detection thresholds of aroma were used. Perhaps recognition thresholds as well as detection thresholds of each compound should be used. The concentration at which panel members can no longer recognize individual characteristic flavor is defined as the recognition threshold. At the recognition threshold level of each citrus oil, Uo of each compound has a special important value. Thus, Uo at the recognition threshold of each oil is expressed by the following equation (2):

$$\text{Limited odor unit (Lo)} = \frac{Tr}{Tc} \tag{2}$$

where: Lo is the limited odor intensity unit of the component, and
Tr is the concentration of individual components at the recognition threshold of volatile oils in ppm, and
Tc is the threshold of the components in ppm.

It is an objective indication that compounds having Lo values more than 1 or nearly equal to 1 make considerable contributions to characteristic aromas. Recognition thresholds of sudachi and lemon oils were measured by the six members of the sensory panel and were found to be 0.033 and 1 ppm, respectively. The Lo values of individual compounds found in lemon oil, which has a recognition threshold of 1 ppm, were calculated. Neral, geranial, and geraniol (Uo > 3.1) had Lo values of approximately 1. However, the components of sudachi oil had Lo values considerably below 1. Key compounds must be found to explain the recognition threshold of 0.033 ppm of sudachi oil. If aroma recognition thresholds in foods can be determined, compounds important in imparting the characteristic flavors can be easily determined because such compounds will have log [Lo] values greater than zero.

Multivariate statistical analysis using logarithmic values of odor units of each component of 6 acid citrus fruits.

Similarity of odor quality of 6 acid citrus fruits was calculated by cluster analysis with the log [Uo] of each component in the oils as shown in Fig. 3. Among the dendrogram patterns of cluster analysis from peak area %, odor unit and log [Uo], dendrogram of log [Uo] (Fig.3) were reflected in the result of sensory evaluation (Table II). Thus, lemon and lime were shown to be most similar, and sudachi, kabosu, yuzu formed another cluster. Kabusu does not seem to be similar to any of the others.

By applying factor analysis and subsequently the VARIMAX procedure for

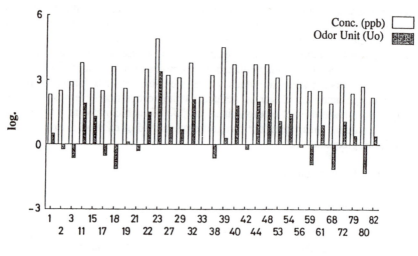

Figure 2. Concentration and odor unit of the oxygenated compounds in sudachi peel oil.

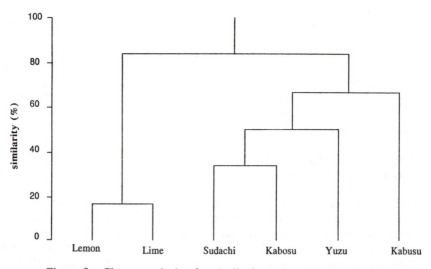

Figure 3. Cluster analysis of peel oils from six acid citrus fruits.

eigenvector rotation, the first two eigenvectors explained 62% of the total variance. Factor analysis of aroma contribution using logarithmic values of odor units of components in 6 acid citrus fruits is shown in Fig. 4. The resultant factor score for the selected compounds (41 compounds) is added on to the result of Factor analysis. The loadings for 6 acid citrus fruits are shown as eigenvectors, the lengths of which roughly indicate their relative importance. The vectors between lemon and lime, and between sudachi and kabosu, indicate a similar direction. Compounds which exist in the same direction of each eigenvector of 6 acid citrus fruits were selected as the most contributing components to the aroma quality. Compounds contributing to individual aroma in 6 acid citrus are listed in Table V. Compounds in sudachi and kabosu selected from factor analysis (Fig. 4) were not always found in both oils. Therefore, compounds contributing to sudachi and kabosu were different. Compounds for lemon and lime were purported to be quite similar, and the balance of the composition between both fruits contributes to the variability in both oils. On the other hand, the difference of the aroma quality of sudachi, yuzu, kabosu and kabusu can be explained by the fact that there are different contributing compounds in each of these fruits.

The differences of aroma quality among 6 acid citrus were attributed to geranial, neryl acetate and neral for lemon and lime; hexanal, cis-limonene oxide and carvone for sudachi; and carvone, 2-nonenal, and cis-limonene oxide for kabosu. Octanal, linalool and thymol were the compounds responsible for the yuzu aroma, and nonanol, linalyl acetate, 2,4-decadienal, and decyl acetate were the contributing compounds to kabusu.

The model sample mixture prepared from compounds having values Log [Uo] > 0 represented the individual aromas of 6 acid citrus fruits. However, we could not find any character impact compounds. Each citrus aroma seems be a combination of a mixture of these compounds identified, or new unknown compounds having large log [Uo] which have not been found. If all compounds having large log [Uo] values were found, each characteristic aroma should be reproduced by a mixture of such compounds.

It was concluded that the GC-sniff method can directly evaluate the aroma quality of one citrus fruit. Log [Uo] values can also indicate contributing components from one sample, and it was not necessary to judge the aroma quality subjectively. Only subjective analysis must be the determination of odor threshold of the components. Multivariate analysis using log [Uo] can deal with several citrus species to evaluate each sample at the same time.

From log [Uo] values it was shown that monoterpene aldehydes, alcohols and their esters, for example, geranial, geraniol, and neral significantly contributed to the characteristic aroma of lemon and lime. Citronellal, citronellol and carvone contributed to sudachi, and citronellol, 2-decenal, geranyl acetate to kabosu. Geraniol, linalyl acetate, geranyl acetate and linalool contributed to kabusu. Unsaturated aliphatic aldehydes contributed especially to yuzu.

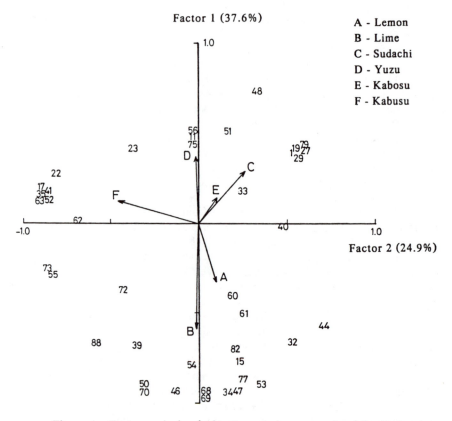

Figure 4. Factor analysis of 41 oxygenated compounds of the peel oils from six acid citrus.

Table V. Compounds Selected by Factor Analysis

No.	Compound	Lemon	Lime	Sudachi	Yuzu	Kabosu	Kabusu
1	hexanal			O			
11	octanal				O		
15	1,8-cineol	O	O				
17	octanol						O
19	cis-linalool oxide			O			
22	nonanal	O	O				
23	linalool				O		
27	cis-limonene oxide			O		O	
29	trans-limonene oxide			O			
33	(E)-2-nonenol					O	
34	borneol	O	O				
35	nonanol						O
41	octyl acetate						O
46	nerol		O				
47	neral	O	O				
48	carvone			O		O	
52	linalyl acetate						O
53	geranial	O					
54	perillaldehyde		O				
55	decanol						O
56	thymol				O		
60	geranyl formate	O					
61	undecanal	O					
62	(E,E)-2,4-decadienal						O
63	nonyl acetate						O
68	citronellyl acetate	O	O				
69	neryl acetate	O	O				
73	decyl acetate						O
75	(E)-2-dodecenal				O		
77	tridecanal	O	O				
79	elemol			O			
82	tetradecanal	O	O				

Acknowledgements

The authors are very grateful to Mr. Kenji Nakanishi and Nobuo Takagi for supplying the acid citrus fruits. We also thank Mr. Hiroyuki Nakatani for his technical assistance. This work was partially supported by a Grant-in-Aid for Scientific Research from the Ministry of Education, Science and Culture of Japan.

Literature Cited
1. Shaw, P. E.; Coleman, R. L. *J. Agric. Food Chem.* **1974**, *22*, 785-787.
2. Wilson, C. W.; Shaw, P. E. *J. Agric. Food Chem.* **1978**, *26*, 1432-1434.
3. Shaw, P. E. *J. Agric. Food Chem.* **1979**, *27*, 246-257.
4. Acree, T. E.; Butts, R. M.; Nelson, R. R.; Lee, C. Y. *Anal. Chem.* **1976**, *48*, 1821-1822.
5. Ullrich, F.; Grosch, W. *Z. Lebensm. Unters. Forsch.* **1987**, *184*, 277-282.
6. Acree, T. E.; Barnard, J.; Cunningham, D. G. *Food Chem.* **1984**, *14*, 273-286.
7. Nitz, S.; In *Topics in Flavour Research*; Berger, R. G.; Nitz, S.; Schreier, P.; Eds.; H. Eichhorn, Marzling-Hangenham, **1985**, pp 43-58.
8. Yang, R-H.; Otsuki, H.; Tamura, H.; Sugisawa, H. *Nippon Nogeikagaku Kaishi;* **1987**, *61*, 1435-1439.
9. Yang, R-H.; Sugisawa, H. *Nippon Shokuhin Kogyo Gakkaishi*, **1990**, *37*, 946-952.
10. Guadagni, D. G.; Okano, S.; Buttery, R. G.; Burr, H. K. *Food Technol.* **1966**. *20*, 166-169.
11. Rothe, M.; Thomas, B. Z. *Lebensm. Unters. Forsch.* **1963**, *119*, 302-310.
12. Shibamoto, T. *J. Food Sci.*, **1986**, *51*, 1098-1099.
13. Buttery, R. G.; Teranishi, R.; Ling, L. C.; Turnbaugh, J. G. *J. Agric. Food Chem.*, **1990**, *38*, 336-340.
14. Sugisawa, H.; Takeda, M.; Yang, R-H., Takagi, N. *Nippon Shokuhin Kogyo Gakkaishi*, **1991**, *38*, 668-674.
15. Yang, R-H.; Sugisawa, H.; Nakatani, K.; Tamura, H.; Takagi, N. *Nippon Shokuhin Kogyo Gakkaishi*, **1992**, *39*, 16-24.

RECEIVED August 19, 1992

Chapter 11

Trace Components in Spearmint Oil and Their Sensory Evaluation

Tomoyuki Tsuneya[1], Masakazu Ishihara[1], Minoru Shiga[1],
Shigeyasu Kawashima[1], Hiroshi Satoh[2], Fumio Yoshida[2],
and Keiichi Yamagishi[2]

[1]Shiono Koryo Kaisha, Ltd., Niitaka 5–17–75, Osaka 532, Japan
[2]Lion Corporation, Honjo 1–3–7, Sumitaku, Tokyo 130, Japan

The basic component in Scotch spearmint oil (*Mentha cardiaca*, Gerard ex Baker) was investigated by means of GC, GC-MS and other analytical techniques. Thirty nine nitrogen compounds were identified as trace components in the oil. Among them, eleven pyridine compounds, including 2-acetyl-4-isopropenylpyridine, were identified for the first time. The chemical structures of these compounds were elucidated by spectral data and by synthesis. The odor qualities of these compounds were evaluated, and a discussion is given of their contributions to the spearmint flavor.

Spearmint oil, along with peppermint oil, is a popular flavoring material used extensively in chewing gums and toothpastes. Although spearmint oil alone is used in these finished products, the practice of mixing it with peppermint oil, known as mixed-mint or doublemint, has enjoyed increasing popularity because it enhances the flavor characteristics of spearmint oil. Taxonomically, spearmint is divided into two species: *Mentha spicata* Huds and *Mentha gentilis* f. *cardiaca*. *Mentha spicata* Huds is a native spearmint, and the latter is Scotch spearmint. Both oils are produced mainly in the USA -- in the Northwestern states of Washington and Oregon, and in the Midwestern states of Michigan, Indiana and Wisconsin. The output of spearmint oil from 1989 to 1991 in the USA is shown in Table 1. Recent growing demands for spearmint flavor and the expansion of planting areas in these producing areas seem to have increased the output of spearmint oil. There is a definite difference in odors between the native spearmint and Scotch spearmint. The Scotch spearmint is generally recognized as superior to the native variety. China is beginning to produce a considerable amount of spearmint oil. Though firm data are not available, it has been predicted that the output of spearmint oil in China could reach about 500 tons.

Canova (*1*) has conducted extensive studies on the flavor components of spearmint oil and has reported 194 compounds. Some characteristic compounds have been reported by Tsuneya et al. (*2*), Takahashi et al. (*3*), Ichimura et al. (*4*),

0097–6156/93/0525–0137$06.50/0

Sakurai et al. (5), Surburg et al. (6), and Shimizu et al. (7). Sakurai et al. (8) have reported three pyridine compounds in spearmint oil. Over 200 flavor components of spearmint oil have been identified (9,10). The authors present here some of the newly identified pyridine compounds characteristic in the oil and sensory evaluations of these compounds.

EXPERIMENTAL PROCEDURE

Sample Preparation: As shown in Figure 1, a 49 kg sample of Midwest Scotch spearmint oil (1983 crop) was extracted three times with 7.5 kg of one normal aqueous solution of hydrochloric acid. The acidic extract was processed to yield 2.8 g of basic components. The GC analysis by flame thermoionic detector (FTD) of the extracted so-called components showed that only 18% of this material were truly composed of basic compounds. Upon further work, it was found that the yield of basic components was 0.00102% (10.2 ppm) of the original oil. Samples were prepared from 400 g amounts of other varieties of spearmint oils and peppermint Mitcham oils (Williamette). The yields of the basic components of Farwest Scotch, Farwest native and peppermint Mitcham oils (Williamette) were 8 ppm, 10 ppm and 5.5 ppm respectively.

Gas Chromatography (GC): A 0.28 mm (i.d.) x 40 m SF-96 fused silica capillary column and 0.25 mm(i.d.) x 60 m DB-1 fused silica capillary column were used in a Hitachi K-163 equipped with flame ionization detector (FID), flame thermoionic detector (FTD) and flame photometric detector (FPD). The oven temperature was programmed from 50° (5 min isothermal) to 240°C at 3°C/min. Similar conditions were used for a DB-1 column. The temperature of the injector was set up at 260°C. Fig. 2 shows the gas chromatogram (FTD) assigned by the peaks of the newly identified basic components. Other peaks assigned are shown in Table 3.

Column Chromatography and Medium Pressure Liquid Chromatography (MPLC): For the purpose of the isolation of unknown compounds, fractionation by column chromatography is shown in Table 2. Further separation of the basic component was undertaken by use of MPLC (Merck Lobar Column:Lichroprep Si-60, 40 - 60 μm, 24 cm x 1 cm i.d.).

Gas Chromatography-Mass Spectrometry (GC-MS): A Hitachi 663 GC was combined with a Hitachi M-80 A mass spectrometer (EI mode) with a M-0101 data processor. A 0.25 mm (i.d.) x 60 m DB-1 fused silica capillary column was used. The oven temperature was programmed from 75°C (5 min isothermal) to 240°C. The temperature of the injector was 250°C. The mass spectra were recorded at an ionization voltage 20 eV at an ion source temperature of 200°C. For high resolution mass spectra (HR-MS), the same instrument was used at an ionization energy of 70 eV.

Infrared Spectrometry (IR): Measurement of the sample was with an IR Jasco IRA-1 instrument.

Proton Nuclear Magnetic Resonance (^{1}H-NMR): Measurement of sample was taken on a Hitachi R-24 B instrument.

Table 1. Output of Spearmint Oil in the USA

Varieties	1989	1990	1991
1 Farwest Scotch	552,000	900,000	1075,500
2 Midwest Scotch	420,000	378,000	680,000
3 Farwest Native	845,000	1,080,000	1,196,000
4 Midwest Native	108,000	86,000	180,000
Total	1,925,000	2,444,000	3,131,500

(Courtesy of I.P. Callison & Sons Incorporated)

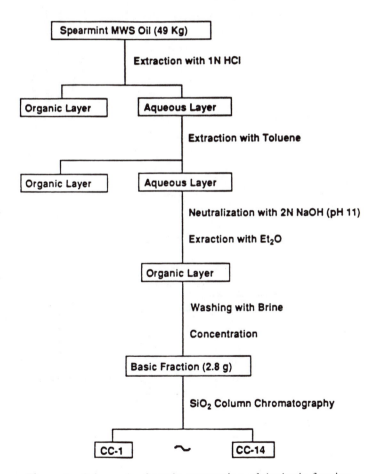

Figure 1. Scheme 1. Sample preparation of the basic fraction.

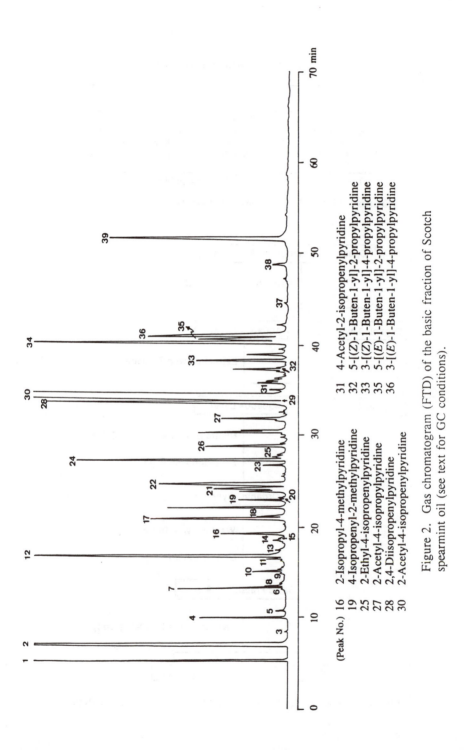

Figure 2. Gas chromatogram (FTD) of the basic fraction of Scotch
spearmint oil (see text for GC conditions).

(Peak No.)	16	2-Isopropyl-4-methylpyridine	31	4-Acetyl-2-isopropenylpyridine
	19	4-Isopropenyl-2-methylpyridine	32	5-[(Z)-1-Buten-1-yl]-2-propylpyridine
	25	2-Ethyl-4-isopropenylpyridine	33	3-[(Z)-1-Buten-1-yl]-4-propylpyridine
	27	2-Acetyl-4-isopropylpyridine	35	5-[(E)-1-Buten-1-yl]-2-propylpyridine
	28	2,4-Diisopropenylpyridine	36	3-[(E)-1-Buten-1-yl]-4-propylpyridine
	30	2-Acetyl-4-isopropenylpyridine		

Table 2. Column Chromatography of the Basic Fraction

Fr. No	Eluting Solvent	Volume (ml)	Yield (mg)
1	hexane	250	0
2	2.5% ether / hexane	200	0
3	5%	75	6
4	5%	75	18
5	10%	75	60
6	10%	75	57
7	20%	75	249
8	20%	75	235
9	30%	80	227
10	30%	70	225
11	50%	100	291
12	70%	100	322
13	ether	150	530
14	methanol	200	521

Table 3. Compounds Identified from the Basic Fraction
in Midwest Scotch Spearmint Oil

Peak No	Compound	Known[a]	Unknown[b]	Means of identification
	(Greater than 10%)[c]			
30	2-Acetyl-4-isopropenylpyridine (X)*	1323	1323	MS, RI, IR, NMR, Syn
2	2-Methylpyridine	770	772	MS, RI
	(1.0 - 10%)			
12	2-Acetylpyridine	991	993	MS, RI
36	3-[(E)-1-Buten-1-yl]-4-propypyridine (XV)*	1438	1442	MS, RI, Syn
22	3-[(Z)-1-Buten-1-yl]pyridine (V)	1131	1131	MS, RI, Syn
24	3-[(E)-1-Buten-1-yl]pyridine (VI)**	1176	1177	MS, RI, Syn
28	2,4-Diisopropenylpyridine (IX)*	1301	1293	MS, RI, IR, NMR, Syn
4	2,6-Dimethylpyridine	846	844	MS, RI
7	3-Ethylpyridine	923	921	MS, RI
17	4-Isopropyl-2-methylpyridine (III)**	1059	1056	MS, RI, Syn
39	5-Phenyl-2-propylpyridine (XVII)	1684	1685	MS, RI, Syn
34	3-Phenylpyridine	1426	1420	MS, RI
1	Pyridine	702	694	MS, RI
	(0.1 - 1.0%)			
27	2-Acetyl-4-isopropylpyridine (VIII)*	1260	1261	MS, RI, Syn
31	4-Acetyl-2-isopropenylpyridine (XI)*	1333	1331	MS, RI, Syn
33	3-[(Z)-1-Buten-1-yl]-4-propylpyridine (XIII)*	1383	1388	MS, RI, Syn
35	5-[(E)-1-Buten-1-yl]-2-propylpyridine (XIV)*	1431	1432	MS, RI, Syn
32	5-[(Z)-1-Buten-1-yl]-2-propylpyridine (XII)*	1378	1379	MS, RI, Syn
22	3-[(Z)-1-Buten-1-yl]pyridine (V)	1131	1131	MS, RI, Syn
18	2-Butylpyridine	1065	1067	MS, RI
20	3-Butylpyridine	1108	1105	MS, RI
21	4-Butylpyridine	1124	1123	MS, RI
6	2,5-Dimethylpyrazine	881	873	MS, RI
25	2-Ethyl-4-isopropenylpyridine (VII)*	1189	1183	MS, RI, Syn
9	2-Ethyl-6-methylpyrazine	960	959	MS, RI
11	5-Ethyl-2-methylpyridine	982	986	MS, RI
5	2-Ethylpyridine	863	866	MS, RI
19	4-Isopropenyl-2-methylpyridine (IV)*	1100	1097	MS, RI
16	2-Isopropyl-4-methylpyridine (II)*	1035	1033	MS, RI, Syn
23	2-Pentylpyridine	1174	1167	MS, RI
38	3-Phenyl-4-propylpyridine (XVI)	1615	1616	MS, RI, Syn
13	3-Propylpyridine	1019	1018	MS, RI
14	4-Propylpyridine	1024	1023	MS, RI
26	Quinoline	1206	1201	MS, RI
	(Less than 0.1%)			
37	3-Benzylpyridine	1500	1497	MS, RI
15	4-Isopropenylpyridine (I)**	1034	1031	MS, RI, Syn
29	Methyl anthranilate	1311	1312	MS, RI
3	3-Methylpyridine	819	820	MS, RI
10	2-Propylpyridine	960	961	MS, RI
8	3-Vinylpyridine	932	928	MS, RI

* newly identified, ** identified first from nature, Known[a]; Kovats Indices (DB-1) of known compound, Unknown[b]; Kovats Indices (DB-1) of unknown compound, (Greater than 10%)[c]; GC area %.

Identification of Components:

Identification of components was accomplished by comparison of known retention indices and mass spectra with those of authentic reference standards. In the absence of published information, authentic samples for mass spectra were purchased. However, the identity of most compounds was confirmed by matching their mass spectra and retention indices with those of synthesized compounds.

Isolation of 2-Acetyl-4-isopropenylpyridine (X): Compound X was isolated by the method mentioned earlier to give 77 mg of a colorless oil: RI 1323; IR (film) 3100, 3070, 1700, 1598, 1250, 910, 853 cm^{-1}; ^{1}H-NMR (CCl$_4$) δ 2.17 (br, s, 3H, CH$_3$) 2.65 (s, 3H, CH$_3$), 5.28(m, 1H, CH$_2$=), 5.61(m, 1H, CH$_2$=), 7.43(dd, J=1.6 and 5 Hz, 1H, ArH), 7.97(d,J=1.6 Hz, ArH) 8.54(d, J=5 Hz, 1H, ArH); EI-MS, m/z(rel. int.) 161(M$^+$, 86), 133(28), 119(100), 118(65), 91(31), 43(30).

Isolation of 2,4-Diisopropenylpyridine (IX): Compound IX was isolated by the method mentioned earlier to give 13 mg as a colorless oil: RI 1293; IR (film) 3100, 1598, 900, 843 cm^{-1}; ^{1}H-NMR (CCl$_4$) δ 2.07 (m, 3H, CH$_3$), 5.16(m, 2H, CH$_2$=) 5.45(m, 1H, CH$_2$), 5.74(m, 1H, CH$_2$), 7.04(dd, J = 1.6 and 5 Hz, 1H, ArH), 7.35(d, J=1.6 Hz, ArH), 8.36(d,J = 5 Hz, 1H, ArH); EI-MS, m/z(rel. int.)159(M$^+$, 100), 158(86), 144(18), 119(25), 91(13).

Preparation of 2-Isopropyl-4-methylpyridine (II):   Compound II was prepared by regioselective addition of isopropylmagnesium bromide to 1-phenoxycarbonyl salt of 4-picoline by applying the method of Comins and Abdulla (*11*) in 44% yield after purification through silica gel chromatography (hexane/ether = 8:2) to give 6.53 g (44%) of II, a colorless oil: RI = 1036; IR(film) 3080, 1603, 1562, 820 cm^{-1}; ^{1}H-NMR (CCl$_4$) δ 1.25(d,J=7 Hz, 6H, two CH$_3$), 2.25(s, 3H, CH$_3$), 2.94(dq, J=7 and 7 Hz, 2H, two CH), 6.76(d, J=5 Hz, 1H, ArH), 6.81(s, 1H, ArH), 8.26(d, J=5 Hz, 1H, ArH); EI-MS, m/z (rel. int.) 135(M$^+$, 40), 134(42) 120(100), 107(28), 93(23), 77(3), 65(5).

Preparation of 4-Isopropenyl-2-methylpyridine (IV) and 2-Ethyl-4-isopropenylpyridine (VII):   Each regioselective alkylation of methyl and ethyl radicals generated from the corresponding carboxylic acid to the 2 position of methyl isonicotinate was performed according to the method of Minisci et al. (*12*). Each of the resulting 2-alkylated esters was treated with methylmagnesium bromide followed by dehydration procedure to give compounds IV and VII in 11% and 18% yields, respectively, after column chromatography on silica gel (ether/hexane = 3:7). Compound IV: RI = 1100; IR (film) 1600, 1534, 900, 835 cm^{-1}; ^{1}H-NMR (CDCl$_3$) δ 2.14(d, J=5 Hz, 1H, ArH), 7.16(s, 1H, ArH), 8.43(d,J=5 Hz, 1H, ArH); EI-MS, m/z(rel. int.) 133(M$^+$, 100), 132(30), 118(19), 117(18), 91(33), 65(11). Compound VII: RI = 1189; IR (film): 1600, 1543, 900, 840 cm^{-1}; ^{1}H-NMR (CDCl$_3$): δ 1.31(t,J=7 Hz, 3H, CH$_3$), 2.11(d,J=1.6 Hz, 3H, CH$_3$), 2.83(q, J=7 Hz, 2H, CH$_2$), 5.18(m, 1H, CH$_2$=), 5.49(m, 1H, CH$_2$=), 7.08(d, J=7 Hz, 5H, ArH), 7.12(s, 1H, ArH), 8.43(d, J=5 Hz, 1H, ArH); EI-MS, m/z(rel. int.) 147(M$^+$, 70), 146(100), 130(5), 119(21), 91(7).

Preparation of 2-Acetyl-4-isopropylpyridine (VIII): Compound VIII was prepared from 2-acetylpyridine and isobutyric acid by applying the method of

Minisci et al. (*12*), and showed the following spectral data: RI=1260; IR (film) 3080, 1700, 1599, 1358, 1202, 843 cm^{-1}; ^{1}H-NMR (CDCl$_3$) δ 1.25(d, J=7 Hz, 6H, two CH$_3$), 2.67(s 3H, CH$_3$, 2.96(dq, J=7 and 7 Hz, 1H, CH), 7.26(dd, J=1.5 and 5 Hz, 1H, ArH), 7.87(d, J=1.5 Hz, 1H, ArH), 8.50(d, J=5 Hz, 1H, ArH); EI-MS m/z(rel.int.) 163(M$^+$, 84), 148(11), 135(28), 121(100), 106(25), 79(10), 43(16).

Preparation of 2,4-Diisopropenylpyridine (IX): An ethereal solution of 3M methylmagnesium bromide was added dropwise to dimethyl 2,4-lutidinate (39 g, 0.2 mol) in ether (400 ml) cooled in an ice water bath. The reaction mixture was stirred for 1h at room temperature and then was refluxed for another 1h. The reaction mixture was poured into a cold sat. NH$_4$Cl. The aqueous layer was extracted three times with a mixed solvent (ether/ethyl acetate = 1:2). The combined organic layer was washed with brine and dried with anhydrous MgSO$_4$. After removal of the solvent, the brownish residue (39.2 g) was chromatographed over silica gel(ether/hexane = 8:2) to give 22.7 g (58%) of the compound (**1**) as colorless crystals and 11.7 g(33%) of a 1:1 mixture of the compound (**2**) and (**3**) as a colorless oil. {Spectral data of compound (**1**)}:IR (KBr): 3280, 2960, 1600, 1180, 1102, 955 cm^{-1};^{1}H-NMR (CDCl$_3$), δ 1.48(s, 6H, two CH$_3$), 1.51(s, 6H, two CH$_3$), 3.69(br.s, 1H, OH), 5.12(br, s, 1H, OH), 7.39(dd, J=1.6 and 5 Hz, 1H, ArH), 7.42(d, J=1.6 Hz, 1H, ArH), 8.25(d, J = 5.0 Hz, 1H); EI-MS, m/z(rel. int.): 195(M$^+$ 0.8), 180(100), 165(12), 137(22). {MS data of compound (**2**)}: EI-MS, m/z, (rel. int.): 179(M$^+$, 100), 164(58), 137(72), 78(35), 59(43), 43(63). {MS data of compound(**3**)}: EI-MS, m/z (rel. int.): 179(M$^+$, 5), 164(100), 121(33), 59(13), 43(5).

The Diol (**1**) (20g, 0.1 mol) was refluxed in conc. H$_2$SO$_4$(50g) and acetic acid (118g) for 1 h. The reaction mixture was cooled and poured into an ice water bath and washed with toluene. The aqueous layer was made basic with 20% NaOH and extracted three times with a mixed solvent (ether/ethyl acetate = 1:1). The organic layer was washed with brine and dried with anhydrous MgSO$_4$. After removal of the solvent, the residue was chromatographed over silica gel (ether/hexane = 1:5) to give 12.5 g compound IX as a colorless oil(yield:78%). The compound IX showed the following spectral data: RI 1301; IR(film):3100, 1598,900,843cm^{-1}; ^{1}H-NMR (CDCl$_3$), δ 2.07(m,3H), 2.17(m,3H), 5.16(m,1H), 5.45(m,1H), 7.04(dd, J=1.6 & 5 Hz, 1H), 7.35(d, J=1.6 Hz, 1H), 8.36(d, J=5 Hz, 1H);EI-MS, m/z(rel. int.):159(M$^+$,100), 158(86), 144(18), 119(25), 91(13).

2-Acetyl-4-isopropenylpyridine (X) and 4-Acetyl-2-isopropenylpyridine (XI): The mixtures of compound (**2**)/(**3**)=1:1(4g,20 mmol), H$_2$SO$_4$ (13g) and glac. acetic acid (30g) were refluxed for 1.5 hr. The reaction mixture was poured into ice water (50 ml) and was extracted with ether/toluene (1:1). The aqueous layer was neutralized with 20% aq. NaOH and was extracted three times with ether. The ether solution was washed with brine and dried (MgSO$_4$). After removal of the solvent, the residue was chromatographed over silica gel (ether/hexane=15:85) to give 1.33 g of compound XI as a colorless oil (yield:37%) and 1.22 g of compound X as colorless oil(yield:34%). Compound X showed the following spectral data: RI 1323: IR (film) 3100, 3070, 1700, 1598, 1250, 910, 853 cm^{-1}; ^{1}H-NMR (CDCl$_3$), δ 2.17(br.s, 3H, CH$_3$), 2.65(s, 3H, CH$_3$), 5.28(m, 1H), 7.43(dd, J=1.6 & 5 Hz, 1H), 7.97(d, J = 1.6 Hz, 1H)

8.54(d, J=5 Hz, 1H); EI-MS, m/z(rel. int.):161(M⁺, 86), 133(28), 119(100), 118(65), 91(31), 43(30). Compound XI showed the following spectral data: RI 1333; IR (film): 3100, 1700, 1590, 1365, 1250, 910, 850 cm⁻¹; ¹H-NMR (CDCl₃), δ 2.24(m, 3H, CH₃), 2.60(s, 3H, CH₃), 5.36(m, 1H), 5.94(m, 1H), 7.56(dd, J=1.6 & 5 Hz, 1H), 7.88(d, J = 1.6 Hz, 1H), 8.74(d, J=5 Hz, 1H): EI-MS, m/z (rel. int.):161(M⁺, 100), 160(70), 146(12), 118(30), 117(20), 91(18), 43(25).

To determine the position of the acetyl group on the pyridine ring, compound X was also prepared by the following selective procedure (S 3 Fig 7). The regioselective acetylation of methyl isonicotinate with paraldehyde was performed by applying the method of Giordano et al. *(13)*. The resulting methyl 2-acetylisonicotinate was transformed to the compound X by Grignard reaction after protecting the acetyl carbonyl group as its ethylenedioxy acetal, followed by dehydration procedure in 23% overall yield. The analytical data of compound X obtained by the above manner were identical with those of the natural one.

5-[(Z)-1-Buten-1-yl)]-2-propylpyridine(XII) and 5-[(E)-1-Butenyl-1-yl)]-2-propylpyridine (XIV): The alkylation of propyl radical to the 6 position of nicotinaldehyde was performed by applying the method of Minisci et al *(12)*. The subsequent Grignard reaction of 6-propylnicotinaldehyde followed by dehydration gave a mixture (3:97) of compound XII and XIV as a colorless oil in 5.8% overall yield after purification through silica gel column chromatography (hexane/ether= 7:3). The minor isomer XII showed the following data: RI 1378; EI-MS m/z(rel. int.)175(M⁺ 5), 174(9), 160(23), 147(100), 132(12), 106(6), 91(3). The major isomer XIV showed the following data: RI 1431; IR(film) 1595, 1560, 1485, 960, 910 cm⁻¹; ¹H-NMR (CDCl₃) δ 0.95(t,J=7 Hz, 3H, CH₃), 1.09(t, J=7 Hz, 3H, CH₃), 1.45-2.50(m, 4H, two CH₂), 2,73(dd, J=7 and 8 Hz, 2H, CH₂), 6.29(m, 2H, CH=CH), 7.05(d, J=7.5 Hz), 1H, ArH), 7.58(dd, J=1.6 and 7.5 Hz), 1H, ArH), 8.45(d, J=1.6 Hz, 1H, ArH); EI-MS m/z(rel.int.) 175(M⁺, 20), 174(15, 160(45), 147(100), 132(40), 92(5).

Preparation of 3-[(Z)-1-Buten-1-yl)]-4-propylpyridine (XIII) and 3-[(E)-1-Butenyl-1-yl)]-4-propyl pyridine (XV): 4-propylnicotinaldehyde was synthesized according to the method of Comins et al. *(14)*. The subsequent Grignard reaction of 4-propylnicotinaldehyde followed by dehydration gave a mixture (3:97) of compound XIII and XV in 23% overall yield after purification through silica gel column chromatography. The minor isomer XIII showed the following data: RI 1383; EI-MS m/z(rel. int.) 175(M⁺, 100) 160(54), 146(33), 132(88), 118(40), 117(35). The major isomer XV showed the following data: RI 1438; IR (film) 3050, 1595, 1465, 1416, 970 cm⁻¹; ¹H-NMR (CDCl₃) δ 0.93(t, J=7 Hz, 3H, CH₃), 1.07(t, J=7 Hz, 3H, CH₃), 1.60(m, 2H, CH₂), 2.17(m, 2H, CH₂), 2.56(dd, J=7 and 8 Hz, 2H, CH₂), 6.04(dt, J=5.5 and 16 Hz, 1H, CH=), 6.45(d, J=16 Hz, 1H, CH=), 6.88(d, J=5 Hz, 1H, ArH), 8.17(d, J=5 Hz, 1H, ArH), 8.43(s, 1H, ArH); EI-MS m/z(rel. int.) 175(M⁺, 83), 160(45), 146(35), 132(100), 131(31), 118(41), 117(39).

Preparation of Other Reference Compounds: 4-Isopropenylpyridine I was prepared by Grignard reaction of methyl isonicotinate followed by dehydration in 63% yield. Spectral data obtained were as follows: RI 1038; IR(film) 3100, 1595, 1402, 905, 830, cm⁻¹; ¹H-NMR(CCl₄) δ 2.08(m, 3H, CH₃), 5.14(m, 1H,

CH$_2$=), 5.45(m, 1H, CH$_2$=), 7.14(m, 2H, ArH), 8.35(m, 2H, ArH),; EI-MS m/z(rel. int.) 119(M$^+$, 100), 118(60), 104(17), 91(33), 79(11).

The mixture (7:93) of Z and E isomers of 3-(1-buten-1-ylpyridine V and VI were prepared by Grignard reaction of nicotinaldehyde followed by dehydration in 63% yield. Spectral data obtained were as follows: Compound V: RI 1131; EI-MS m/z(rel. int.) 133(M$^+$, 100), 132(43), 118(85), 117(43), 91(23), 65(13). Compound VI: RI 1176; IR(film)3050, 1570, 1420, 1025, 970, 800, 710cm^{-1}; ^{1}H-NMR (CCl$_4$) δ 1.07(t, J=7 Hz, 3H, CH$_3$), 2.21(m, 2H, CH$_2$), 6.17(m, 2H, CH=CH), 6.98(dd, J=5 and 8 Hz, 1H, ArH), 7.44(dt, J=1.6 and 8Hz, 1H, ArH), 8.21(dd, J=1.6 and 5 Hz, 1H, ArH); EI-MS m/z(rel. int) 133(M$^+$, 85), 132(45), 118(100), 117(47), 91(24), 65(11).

4-Isopropyl-2-methylpyridine (III) was prepared according to the method of Comins et al. (*11*) in 52% yield. Spectral data obtained were as follows: RI 1060; IR (film) 3080, 3030, 2975, 1603, 1560, 920, 830 cm^{-1}; ^{1}H-NMR (CDCl$_3$) δ 1.20(d, J=7 Hz, 6H, two CH$_3$), 2.47(s, 3H, CH$_3$), 2.79(dq, J=7 and 7 Hz, 1H, CH), 6.84(d, J=5 Hz, 1H, ArH), 6.87(s, 1H, ArH), 8.25(d, J=5 Hz, 1H, ArH); EI-MS m/z(rel. int.) 135(M$^+$, 64), 120(100), 106(4), 93(6), 77(13), 65(5), 42(6), 41(6).

3-Phenyl-4-propylpyridine(XVI) was prepared from 3-phenylpyridine and butyric acid by applying the method of Minisci et al. (*12*) and it showed the same physiochemical properties of those reported by Sakurai et al. (*8*).

5-Phenyl-2-propylpyridine(XVII) was prepared according to the method of Sakurai et al. (*8*) and it showed the same physiochemical properties as those reported.

Results and Discussion

Identification of Pyridine Compounds: Fig. 2 shows the gas chromatogram of the basic part in Midwest Scotch spearmint oil, and also shows the peaks of the eleven new pyridine compounds identified in this study. As can be seen in Figure 2, the existence of more than 100 nitrogen compounds was detected first with FTD and subsequently they were analyzed by GC-MS. It had been predicted that many of these compounds might be unknown or unfamiliar compounds. This fact led to much interest in and elucidation of the structures. In particular, the main component which showed a unique mass spectrum was especially interesting. The main component had a molecular weight of 161 m/z. The GC-MS (Fig. 3) and high resolution mass spectrum showed the molecular formula, C$_{10}$H$_{11}$NO; [M$^+$] 161.0834. Compound X isolated by chromatography was submitted to the following instrumental analyses. IR spectrum (Fig. 4) shows the presence of an acetyl carbonyl group at 1700 and 1250 cm^{-1}. The ^{1}H-NMR signals (Fig. 5) at δ 2.17 (br, s, 3H), 5.28(m,1H) and 5.61(m,1H) suggested the existence of an isopropenyl group. The signals at δ 7.43(dd, J=1.6 and 5 Hz, 1H), 7.97(d, J=1.6 Hz, 1H) and 8.54(d, J=5 Hz, 1H) suggested that the acetyl and the isopropenyl groups are attached to the 2- and 4-positions or the reverse positions on the pyridine ring, respectively. These spectral data supported two assigned structures of 2-acetyl-4-isopropenyl pyridine (X) or 4-acetyl-2-isopropenylpyridine (XI). The structure of compound X was determined first by synthesis as shown

HR-MS Found [M⁺] *m/z* 161.0834 $C_{10}H_{11}O_1N_1$ RI=1323

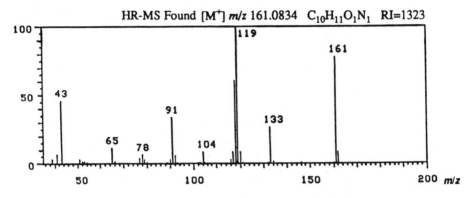

Figure 3.   Mass spectrum of compound **X** (isolated).

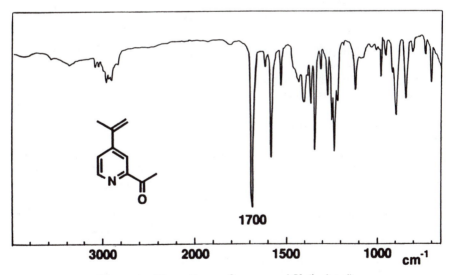

Figure 4.   IR spectrum of compound **X** (isolated).

in Scheme 2, (Figure 6), and the position of the substituted group was confirmed by selective synthesis as shown in Scheme 3 (Figure 7). The spectral properties of the synthetic product were identical with those of one isolated from spearmint oil. Small amounts of 4-acetyl-2-isopropenylpyridine (XI) were also confirmed by comparing its mass spectrum and RI data with those found in nature. On the other hand, compound IX (Peak No. 28 in Fig. 2) which appeared just ahead of the main component in GC showed a molecular weight of 159 m/z from GC-MS and the molecular formula $C_{11}H_{13}N$ from the HR-MS (Fig. 8). The IR spectrum (Fig. 9) showed no absorption of the carbonyl group. The ^{1}H-NMR spectrum (Fig. 10) showed the signals at 2.07 (m, 3H, CH_3), 2.17 (m, 3H, CH_3), 5.16 (m, 2H, $CH_2=$), 5.45 (m, 1H, $CH_2=$) and 5.74 (m, 1H, $CH_2=$), which suggest the existence of two isopropenyl groups. The coupling patterns of three proton signals of the pyridine ring were quite similar to those of compound X. These spectral data supported the assigned structure of 2,4-diisopropenylpyridine (IX), which was confirmed by the synthesis drawn in scheme 2. The spectral data of the synthesized compound were absolutely identical with those of the natural one.

Besides these compounds, unknown compounds were further studied to elucidate their chemical structures. To determine the structures of other compounds for which authentic MS data were not available, many pyridine compounds had to be synthesized to elucidate their structures by examining their mass spectra. By synthesizing the unknown compounds and by analyzing their mass spectra and by matching the RI with those of the natural ones, thirty nine pyridine compounds were identified as shown in Table 3. In this table, the compounds with an asterisk marks are newly identified ones. Double asterisk marks indicate that the compounds which were identified for the first time in nature but had been previously synthesized. Figures 11 to 17 show mass spectra of new pyridine compounds, except for compounds IX and X. The mass spectra of XIV and XV are similar to those of compounds XII and XIII because they are the trans and cis forms, respectively, see Figures 16 and 17.

Existence of Pyridine Compounds in Nature

Maga (*15*) and Vernin (*16*) have reviewed in detail pyridine compounds identified in nature, in various natural products such as fish, meat, poultry, vegetables, cereals, nuts, dairy products, fruits, spices, non-alcoholic beverages like tea and coffee, alcoholic beverages, tobacco, and essential oils. Approximately 100 pyridine derivatives with various ring substitutions have been identified. In essential oils, some characteristic pyridine compounds have been reported in jasmin (*17*), orange flower (*18*), and peppermint (*8*). Toyoda (*17*) reported 14 unique pyridine compounds such as esters of nicotinic acid from jasmine absolute. Sakurai (*8*) identified in peppermint and spearmint oils several pyridine derivatives substituted at 3,4- and 2,5-positions with phenyl groups. It is also noticed that 16 and 30 nitrogen compounds have been identified in black tea (*19*) and Burley tobacco (*16*), respectively. One of the interesting features of this work is that 8 pyridine compounds have been reported which have isopropenyl, isopropyl, and acetyl groups substituted in the 2,4-positions (Figure 18). As for the few 2,4-substituted pyridines found in nature so far, 2,4-dimethylpyridine in Burley

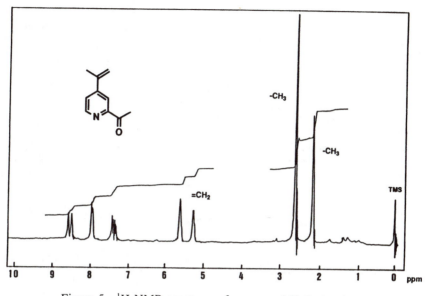

Figure 5. ¹H-NMR spectrum of compound **X** (isolated).

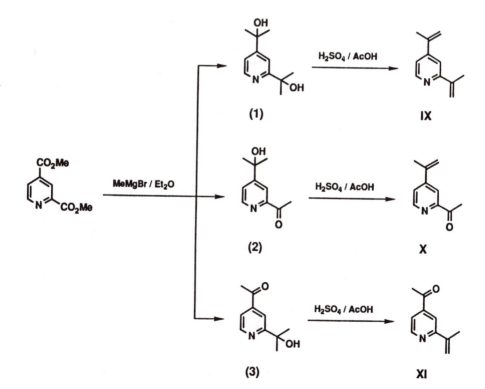

Figure 6. Scheme 2. Synthesis of compounds **IX**, **X**, and **XI**.

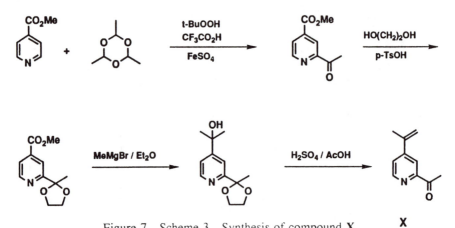

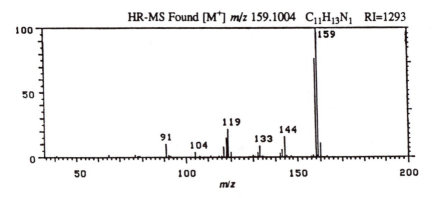

Figure 7. Scheme 3. Synthesis of compound **X**.   **X**

HR-MS Found [M$^+$] m/z 159.1004 C$_{11}$H$_{13}$N$_1$ RI=1293

Figure 8. Mass spectrum of compound **IX** (isolated).

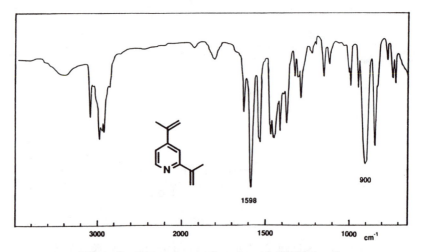

Figure 9. IR spectrum of compound **IX** (isolated).

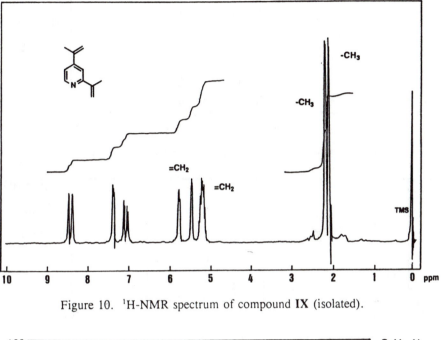

Figure 10. ¹H-NMR spectrum of compound **IX** (isolated).

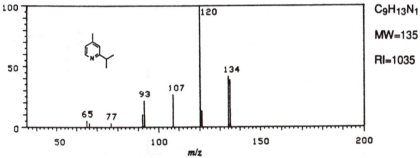

Figure 11. Mass spectrum of compound **II** (synthesized).

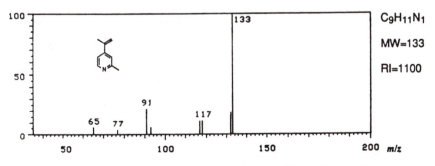

Figure 12. Mass spectrum of compound **IV** (synthesized).

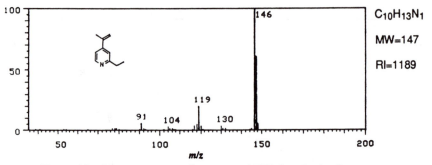

Figure 13.   Mass spectrum of compound **VII** (synthesized).

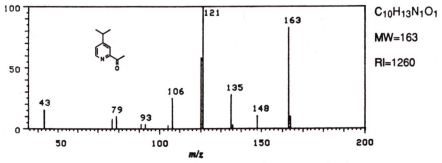

Figure 14.   Mass spectrum of compound **VIII** (synthesized).

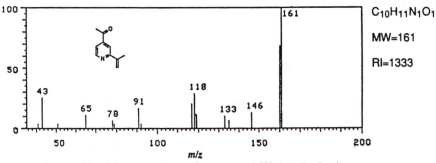

Figure 15.   Mass spectrum of compound **XI** (synthesized).

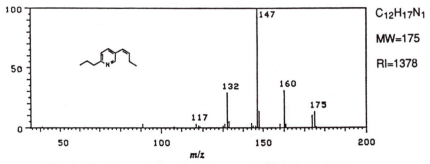

Figure 16.   Mass spectrum of compound **XII** (synthesized).

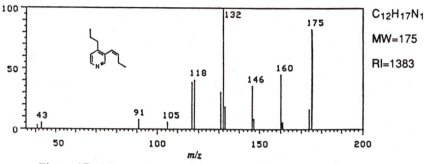

$C_{12}H_{17}N_1$

MW=175

RI=1383

Figure 17. Mass spectrum of compound **XIII** (synthesized).

I

II

III

IV

V

VI

VII

VIII

IX

X

XI

XII

XIII

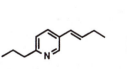

XIV

XV

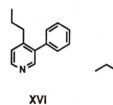

XVI

XVII

Figure 18. Pyridine compounds in spearmint oil.

tobacco (20) and four 2,4-pyridine compounds, 2-isobutyl-4-methylpyridine, 2-isobutenyl-4-methylpyridine,2-isobutanoyl-4-pyridine,and2-(1-hydroxy-2-methyl-1-propyl)pyridine, in fig absolute (21) have been reported.

     Table 4 lists the major pyridine compounds identified from Midwest Scotch, Far West Scotch, Far West native spearmint oil and Mitcham peppermint oil (Williamette). Fig. 18 shows characteristic pyridine compounds identified from spearmint oil. As can be seen in Table 4, the main component is 2-acetyl-4-isopropenylpyridine in each of the spearmint oils, and the other pyridine compounds which exist in Midwest Scotch are not so different qualitatively when compared with Far West Scotch and native. These three oils have some qualitative differences. In particular, the differences in quality between Scotch and native are quite clear. From the information in this table, it can be observed that the contribution of flavor in the basic components of the individual oil does not seem to determine the organoleptic differences. On the other hand, analysis of the basic component of peppermint oil showed that it compared well with spearmint oil. Although spearmint and peppermint oils have pyridine compounds common to both, there are some differences. Compounds VIII, IX, X and XI are found in spearmint oil but not in peppermint oil. The fact that those compounds are found only in spearmint oil prompted the study of spearmint flavor.

Sensory Evaluation of Pyridine Compounds

Table 5 shows the odor descriptions evaluated by seven flavorists. The individual compounds were evaluated at the concentration of 5% in alcohol. Of these compounds, 2-Acetyl-4-isopropenylpyridine, which is unique to spearmint and also is the main component in the oil, has a grassy-sweet, minty, somewhat amber-like odor. The terms used in evaluating pyridine compounds have been reported by Maga (13), Vernin (14), Winter et al. (22), Buttery (23), Suyama et al. (24). These terms are green, astringent, bitter earth or burnt note. Acetylpyridine derivatives are considered to have roasted and coffee-like odor. The odor descriptions in Table 5 are green earth, and roasted or brownish, though the characters such as aged, fermented, herbal, cinnamate and somewhat animalic odor are added to the terms already mentioned. Harsh terms such as astringent disappeared in the evaluation of these compounds, and somewhat more benign terms have appeared.

     Table 6 shows the sensory evaluation of some pyridine compounds associated with spearmint flavor. The standard spearmint flavor is composed of 30 flavor chemicals. To this standard spearmint flavor, the following pyridine compounds were added at ppb levels to evaluate their impact. These compounds were added to improve the spearmint flavor by giving a grassy-sweet note to this spearmint flavor which had a somewhat fishy odor. When 2-acetyl-4-isopenyl-pyridine was added at the concentration of 40 ppb to the standard spearmint flavor, it imparted a slightly roasted, fermented or aged odor, augmenting a grassy-sweetness, accentuating the sweetness of a terpene top note. On the other hand, 2,4-diisopropenylpyridine and 2-acetyl-4-isopropylpyridine gave characteristic properties to the standard one at the concentration of 2 and 8 ppb, respectively, as listed in Table 6. Further, when 2-acetyl-4-isopropenylpyridine

Table 4. Major Pyridine Compounds in Spearmint and Peppermint Oil (ppm)

Compound	A	B	C	D
Pyridine	0.98	0.12	0.46	0.26
2-Methylpyridine	1.25	0.06	t	0.02
2,5-Dimethylpyrazine	0.06	0.11	0.58	0.02
2-Acetylpyridine	0.20	0.14	0.13	0.07
2-Isopropyl-4-methylpyridine (II)	0.05	0.05	0.13	0.34
4-Isopropyl-2-methylpyridine (III)	0.12	0.11	0.14	1.90
4-Isopropenyl-2-methylpyridine (IV)	0.05	0.02	0.02	t
3-[(Z)-1-Buten-1-yl]pyridine (V)	0.10	0.09	0.23	0.15
3-[(E)-1-Buten-1-yl]pyridine (VI)	0.20	0.17	0.59	0.22
2-Ethyl-4-isopropenylpyridine (VII)	0.04	0.03	0.14	t
Quinoline	0.09	0.11	0.11	0.06
2-Acetyl-4-isopropylpyridine (VIII)	0.06	0.02	0.04	-
2,4-Diisopropenylpyridine (IX)	0.35	0.26	0.40	-
2-Acetyl-4-isopropenylpyridine (X)	3.34	3.26	3.54	-
4-Acetyl-2-isopropenylpyridine (XI)	0.05	t	0.03	-
5-[(Z)-1-Buten-1-yl]-2-propylpyridine (XII)	0.02	0.02	0.02	0.02
3-[(Z)-1-Buten-1-yl]-4-propylpyridine (XIII)	0.14	0.25	0.49	0.01
3-Phenylpyridine	0.58	0.34	0.57	0.41
5-[(E)-1-Buten-1-yl]-2-propylpyridine (XIV)	0.08	0.34	0.14	0.10
3-[(E)-1-Buten-1-yl]-4-propylpyridne (XV)	0.26	0.44	0.05	t
3-Phenyl-4-propylpyridine (XVI)	0.05	0.03	0.07	0.04
5-Phenyl-2-propylpyrdine (XVII)	0.28	0.18	0.22	0.12

A: Spearmint Midwest Scotch, B: Spearmint Farwest Scotch, C: Spearmint Farwest Native,
D: Peppermint Mitcham(Willamette)

was added to toothpaste material together with the standard one, in addition to these evaluations in Table 6, rough properties of the artificial flavor seemed to be smoothed out. It is interesting that these compounds in such low concentrations can contribute towards approximating natural flavor. Other compounds of low threshold values have also been known to cause similar effects. However, these compounds alone should not be considered as representing the exclusive role of the basic component of spearmint oil. Besides this compound, a combination of the components with some other characteristic minor pyridine components had an impact towards approaching a "natural" flavor. In general, a pyridine component has not been evaluated to be as good a contributor as a pyrazine component because pyridine compounds are not as palatable as pyrazine compounds.

It should be emphasized that pyridine compounds can enhance natural flavors. They can modify the flavor quality of the synthetic to resemble the natural one. The study on pyridine components in nature will be continued.

Genetic Consideration: As can be seen from Table 4, a feature of the basic components of spearmint oil is the existence of the 2,4-disubstituted pyridine derivatives which are rare in nature. These compounds are not found in peppermint oil. A possible mechanism of their characteristic skeleton can be explained as shown in Scheme 4 (Figure 19). The oxidative degradation product

Table 5. Odor Description of Synthesized Pyridine Compounds

Compound	Odor description
4-Isopropenylpyridine (I)	green-bitter, nutty-beany, slightly sweet
4-Isopropenyl-2-methylpyridine (IV)	ether like, browny-acidy, radish(ozone like)
2-Ethyl-4-isopropenylpyridine (VII)	slightly nutty, herbal, bitter
2,4-Diisopropenylpyridine (IX)	earthy, slightly seaweed, somewhat citrus
2-Isopropyl-4-methylpyridine (II)	earthy green, somewhat sour and citrus
4-Isopropyl-2-methylpyridine (III)	amine like, ozonous green, violet-perilla
3-[(Z&E)-1-Buten-1-yl]pyridine (V & VI)	herbal, white floral like, minty
5-[(Z&E)-1-Buten-1-yl]-2-propylpyridine (XII & XIV)	somewhat rose, fermented beany, wormwood
3-[(Z&E)-1-Buten-1-yl]-4-propylpyridine (XIII & XV)	earthy green, green beany, powdery musk like
3-Phenylpyridine	nutty, roasted soybean, methyl cinnamate like
3-Phenyl-4-propylpyridine (XVI)	minty, sweet, fermented earthy
5-Phenyl-2-propylpyridine (XVII)	green tomato leaf, slightly methyl cinnamate like
2-Acetyl-4-isopropenylpyridine (X)	grassy-sweet, minty, somewhat amber like
4-Acetyl-2-isopropenylpyridine (XI)	weak herbal green, fermented roast
2-Acetyl-4-isopropylpyridine (VIII)	grassy-green leaf, green herbal, somewhat violet

Table 6. Sensory Evaluation of Pyridine Compounds to Spearmint Flavor

Compound	Concentration	Odor Description
2-Acetyl-4-isopropenylpyridine (X)	40 ppb	slightly roasted and brownish, fermented odor, aged, soft, pull out the sweetness of terpene, augment the bottom note
2,4-Diisopropenylpyridine (IX)	2 ppb	like hay with sweetess, slightly brownish, round out, gorgeous
2-Acetyl-4-isopropylpyridine (VIII)	8 ppb	enhancing thickness, sweetness peculiar to spearmint, naturality

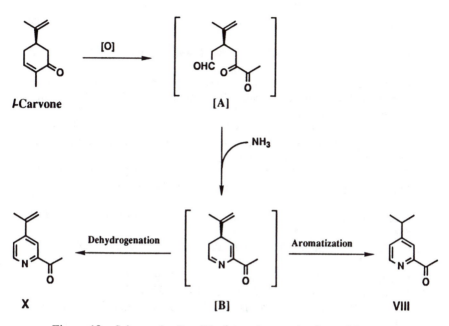

l-Carvone   [A]

X   [B]   VIII

Figure 19. Scheme 4. Possible formation mechanism of 2-acetyl-4-isopropenylpyridine.

of 1-carvone, which is a main component of spearmint oil, reacts with ammonia to give dihydropyridine intermediate B through A. Aromatization of B affords 2-acetyl-4-isopropylpyridine VIII. On the other hand, dehydrogenation of B produces 2-acetyl-4-isopropenylpyridine X. If pyridine compounds were proved to be formed genetically from terpene compounds in this way, it could be said that it is quite rare in nature. The formation of the genetic mechanism of pyridine compounds common in spearmint and peppermint oil is another topic of interest. Study on this matter will be continued.

Literature Cited:

1. Canova, L. The composition of Scotch spearmint oil. In *5th International Congress of Essential Oils,* Abstract paper, QT/b-22, Brazil, An. Acad. Bras. Cienc., **1971**, pp 273-277.
2. Tsuneya, T.; Yoshioka, A.; Shibai, T.; Shiga, M. *Koryo,* **1973**, *104,* 23-26.
3. Takahashi, K.; Muraki, S.; Yoshida, T. *Agric. Biol. Chem.,* **1981**, *45,* 129-132.
4. Ichimura, N.; Matsura, Y.; Kato, Y., *25th Terpene, Essential Oils and Aroma Congress,* **1981**, 18-20.
5. Sakurai, K.; Takahashi, K.; Yoshida, T. *Agric. Biol. Chem.,* **1983**, *47,* 1249-1256.
6. Surburg, H.; Kopsel, M. *Flavor and Fragrance Journal,* **1989**, *4,* 143-147.

7. Shimizu, S.; Shibata, H.; Karasawa, D., Kozaki, T. *J. Ess. Oil Res.*, **1990**, *2*, 81-86.
8. Sakurai, K.; Takahashi, K.; Yoshida, T. *Agric. Biol. Chem.*, **1983**, *47*, 2307-2317.
9. Lawrence, B. M. Monoterpene interrelationship in the mentha genus: a biogenetic discussion. In *Essential Oils*, Mookherjeem B. D.; Mussinan, D. J., Eds., Allured, Wheaton, Ill., **1981**, pp 1-81.
10. Maarse, H.; Visscher, C. A.; Willemsens, L. C. *Volatile Compounds in Foods -- Qualitative and Quantitative*, TNO-CIVO Food Analysis Institute, Zeist, The Netherlands, **1989**, pp 110-136.
11. Comins, D. L.; Abdullah, A. H. *J. Org. Chem.*, **1982**, *47*, 4315-4319.
12. Minisci, F.; Bernardi, R.; Bertini, F.; Galli, R; Perchinummo, M., *Tetrahedron*, **1971**, *27*, 3575-3579.
13. Giordano, C; Minisci, F.; Vismara, E.; Levi, S. *J. Org. Chem.*, **1986**, *51*, 536-537.
14. Comins, D. L.; Smith, R. K.; Stroud, E. D. *Heterocylces*, **1984**, *22*, 339-344.
15. Maga, J. A. *J Agric. Food Chem.*, **1981**, *29*, 895-898.
16. Vernin, G. *Perfumery and Flavorist*, **1982**, *7*, 23-35.
17. Toyoda, T; Muraki, S.; Yoshida, T. *Agric. Biol. Chem.*, **1978**, *42*, 1901-1905.
18. Sakurai, K.; Toyoda, T.; Muraki, S.; Yoshida, T. *Agric. Biol. Chem,*, **1979**, *43*, 195-197.
19. Vitzhum, O. B.; Werkhoff, P.; Hubert, P. *J. Agric. Food Chem.*, **1975**, *23*, 999-1003.
20. Neurathe,G. B. *Beitr. Tabakforsch.*, **1969**, *5*, 515-518.
21. Kaiser, R. New natural products of structural and olfactory interest identified in fig leaf absolute (*Ficus carica* L.). In *Prog. Essent. Oil. Res.*, Walter de Gruyter & Co., Berlin, **1986**, pp 227-239.
22. Winter, M.; Gautschi, F.; Flament, I; Stoll, M.; Goldman, I. M. **1975**, *U. S. Patent* 3,900,582. **1972**, *U. S. Patent* 3,702,253, Flament, I.; Stoll, M.; **1976**, *U. S. Patent* 3,931,246; **1976**, *U. S. Patent* 3,931,245.
23. Buttery, R. G.; Ling, L. C.; Teranishi, R.; Mon, T. R. *J. Agric. Food Chem.*, **1977**, *25*, 1227-1229.
24. Suyama, K.; Adachi, S. *J. Agric. Food Chem.*, **1980**, *28*, 546-549.

RECEIVED September 28, 1992

Chapter 12

Therapeutic Properties of Essential Oils and Fragrances

G. Buchbauer[1], W. Jäger[1], L. Jirovetz[1], J. Ilmberger[2], and H. Dietrich[3]

[1]Institute of Pharmaceutical Chemistry, University of Vienna,
Währingerstreet 10, Vienna A–1090, Austria
[2]Clinic of Physical Medicine, Klinikum Grosshadern, Marchionistreet 15,
D–8000, Munich 70, Germany
[3]Central Laboratory Animal Facilities, University of Innsbruck, Fritz-Pregl
Street 3, A–6020, Innsbruck, Austria

As a continuation of our studies on the biological (mainly sedative) properties of essential oils and fragrance compounds, new results in aromatherapy research will be presented. After a short introduction dealing with therapeutic properties of essential oils and fragrances in general, emphasis will be on the correct definition of the term "aromatherapy". Examples of how distinct biological actions evoked only by means of inhalation of essential oils and fragrance compounds will be presented.

Therapeutic applications of essential oils and fragrance compounds have been known since ancient times (1-3). The fumigation of sick persons surely is one of the most ancient treatments mentioned in the epic poem "Gilgamesch". Though the earliest uses of fumigation involved religious rites, this practice introduced the concept of disinfection, cleaning and refreshing the air surrounding a sick person. It is known that King Salomo and the Egyptian Queen Cleopatra used pillows filled with dried rose petals to facilitate sleep.

The therapeutic use of fragrances and essential oils is well documented from ancient times through the Middle Ages until the 19th century. Single chemicals as medication as we have today were not known. Only natural remedies were used. The disinfectant effects of fumes from smelling candles, or fragrance candles as they would be called today, were used very often in those days. In order to banish the bad, stale air, the Roman writer Plinius recommended that fresh peppermint plants be hung in sick rooms. Also doctors tried to protect themselves against infectious air by sniffing essential oils via an artificial beak (One can imagine this as a primitive form of a protective cloth). They disinfected their hands by clasping the knob of a stick impregnated with fragrance compounds. The pomander, a short form of the French "pomme d'ambre", is a perforated little ball- or apple-shaped container worn like a necklace and filled with a mixture of spices and fragrances.

0097–6156/93/0525–0159$06.00/0

"Sachet" is a little linen bag filed with dried herbs or other fragrant materials, such as rose petals, lavender flowers, or hay blossoms. In recent years, there seems to be a renaissance in the use of these fragrance filled sachets.

Therapeutic properties of fragrance compounds in general

After this introduction and historical stroll into the wide field of therapeutic usage of essential oils and fragrances, some recent application of fragrance compounds should be discussed (*1-3*). At first the antibacterial activity of many fragrances and essential oils must be mentioned (*4*). Many authors and scientists throughout the world have published on this subject (*5-7*).

Aromatic substances are often used to treat coughs and unspecific irritations of the respiratory tract. The fragrance compounds are inhaled and reach the mucosa of the respiratory tract. Their efficacy is felt even in the finest bronchioles without any increase of bronchiomotoric activity. Cough impulses are reduced by the antiphlogistic and spasmolytic effects of these aromatic compounds. The antibacterial activity has already been mentioned.

Another application of fragrance compounds is as an agent to alleviate the symptoms of cacosmia, a special form of parosmia. This condition is caused either by tumors in the nose or by a traumatic destruction of the olfactory nervous system, causing those suffering from this affliction to smell only unpleasant odors.

A modern use of fragrance compounds is their use in the control, regulation, or suppression of the appetite of overweight persons. Unpleasant odors reduce the enjoyment of eating and reduce appetite. The use of appealing odors to make material of low food value has been reported some years ago (*8*). This use of fragrance compounds to control or suppress appetites seems striking in its irony when many citizens of the third world face hunger and starvation.

The application of fragrance compounds to stimulate or tranquillize the nervous system constitutes a large field. Smelling salts are often used to revive people who have fainted. Many people experience a pleasant feeling after breathing certain fragrances. This quasi thymoleptic activity can be used to influence the psychic state of people, e.g. to enhance their enjoyment of work and hence their working efficiency, to temper a bad mood, or to alleviate fatigue. Even anxiolytic properties (e.g. against phobias) can be attributed to fragrance compounds and perfumes (*9*). They can stimulate or tranquillize the cortex. Many fragrances show sedative, spasmolytic, and tranquillizing properties and are used in aromatherapy to facilitate sleep, relieve nervous tension, or contribute to a calm and serene attitude. Fragrance candles and sleeping pillows are used for these purposes.

Aromatherapy

Regarding all aforementioned applications of fragrance compounds and essential oils, one can easily discern that all of the effects are evoked only by inhalation. And this observation leads directly to the correct definition of the term "aromatherapy" (*1-3*):

Aromatherapy is the therapeutic use of fragrances or at least of mere volatiles to cure or to mitigate or to prevent diseases, infections, and indispositions only by means of inhalation.

Aromatherapy is not the application of medication as aerosols, incorporation by ways other than inhalation, such as massage or any cosmetic treatment, combination with cosmic, magnetic, astrological or anthroposophic phenomena.

The therm "aromatherapy" was coined by the French cosmetic chemist Gattefossé in the thirties emphasizing the good antibacterial properties of essential oils and their excellent skin permeability (*10*). But these two points are dermatological-cosmetic considerations, and all of his successors gave way to a great error, to a great misunderstanding because aromatherapy has to do something with aroma, and aroma alone! Aroma is a fragrance, a sweet smell, a subtle pervasive quality! Can a smell be eaten? Can a fragrance be rubbed into the skin? Can a subtle pervasive quality be massaged on skin areas? Perhaps an essence or a solution of fragrances or an essential oil can be, but not an aroma. Even if many persons may try to argue with the author of this paper, they cannot contradict the concise definitions in the Oxford dictionary, a very reliable source (*11*).

Scientific examples

One of the first studies dealing with aromatherapy, in the correct sense, was the investigation of the sedative properties of hop pillows (*12*) used for many years in European folk medicine against difficulties in falling asleep. By means of headspace chromatography of the air above such a pillow, it was possible to detect the three main volatile hop constituents: acetone, myrcene, and dimethylvinylcarbinol. This alcohol was shown to be a sedative in a pharmacological experiment with mice.

Another leading study has been performed by Kovar et al. (*13*). They investigated the efficiency of the essential oil of rosemary and its main constituent, 1,8-cineole. The medicinal folk usage of this oil as a remedy against exhaustion was shown to be justified by an experiment in which this essential oil was given to mice to inhale. The locomotoric activity of the test animals increased significantly when they inhaled this oil.

Own results

In our research we focussed our interest in studying the sedative and tranquillizing properties of a number of essential oils and single fragrance compounds in an aromatherapeutical sense. We were able to show a distinct decrease in the motility of mice by simply making them inhale various materials.

At first we investigated the essential oil of lavender (*Lavandula angustifolia* Mill.) because the oil, as well as the flowers, from this plant have been used from ancient times in folk medicine as a sedative and tranquillizer (e.g. lavender filled sleeping pillows and lavender fragrance candles). Lavender oil and its main constituents, linalool and linalyl acetate, reduced the motor activity of the test animals about 80%, see Figure 1 (*14*). We were able to reduce induced motor activity remarkably. Prior to exposure of mice to the fragrance material, we administered caffeine intraperitoneally, inducing hyperagitation to about 60% of normal. Lavender oil, and to a lesser degree its main constituents, reduced the induced nervous state almost quantitatively.

In the same manner, we tested the essential oil of East Indian sandalwood. It is claimed that this highly prized essential oil exerts a calming effect, shown by a CNV curve (*15*). This contingent negative variation is an upward shift of the brain waves which can be observed in an EEG of a test person who is expecting something to happen. In animal studies, we were able to indeed observe a decrease in the motility of the test mice. This observation is in agreement with the studies with human beings by the Japanese scientists (*15*). However, it must be noted that in the case of sandalwood, the amount of volatiles had to be increased in order to detect a noticeable sedative effect. This can be explained by the fact that the sesquiterpene alcohols, α- and β-santalol, the two main constituents of this essential oil, have very low vapor pressure.

In a large study (Buchbauer, G; Jirovetz, L.; Jäger, W.; Plank, Ch.; Dietrich, H., paper submitted, 1992), we investigated about 40 fragrances, among them 9 essential oils, to observe their ability to decrease the motility of the test animals. The greatest decrease of motor activity (more than 60%) was observed after the exposure to the already mentioned lavender oil, followed by linalool, linalyl acetate, and neroli oil. Seven fragrance compounds ranked behind with a decreasing rate between 30 to 50%, e.g. sandalwood oil (already mentioned), benzaldehyde, α-terpineol, citronellal, and β-phenyl-ethyl acetate. Rose oil, which is credited as having a calming, soothing, and relaxing power, astonishingly did not show a significant decrease of motility (a value of less that 10%). Other compounds even caused an increase in motility, like isoborneol (increase of 50%), or a mixture of orange terpenes (increase of 35%).

Lavender oil, linalool, linalyl acetate, and astonishingly, isoeugenol (which increased the motor activity of test mice under normal conditions), are the best "sedatives" (decreasing agitation by more than 50%) to decrease caffeine-induced overagitation. These materials are followed by 19 fragrance substances, among them, carvone, benzaldehyde, citronellal, eugenol, maltol, and β-phenyl ethanol decreased overagitation by 11 to 49%. Again, rose oil proved ineffective.

Finally, some data of our first human experiments are presented. Subjects had to perform computer-based reaction-time tests while breathing pure air or inhaling lavender oil vapors. The two tests, which are parts of a larger computerized attention battery (Munich Attention Tests [MAT], developed by Ilmberger, J., and Karamat, E., Universities Munich/Innsbruck, personal communication, 1992), consisted of visual single reaction time measurements without and with a warning tone.

Subjects had to press a button with the index finger until the appearance of a large filled circle on the computer screen. Upon appearance of the stimulus, the button had to be released, and a second button had to be pressed as quickly as

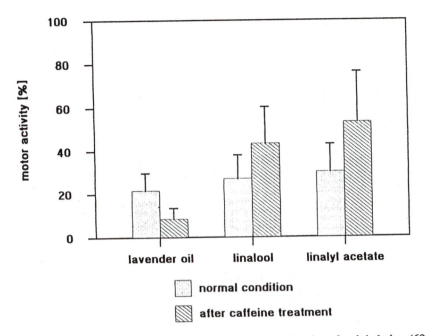

Figure 1.   Decrease of motor activity of 6-8 week old mice after inhalation (60 minutes) of lavender oil, linalool, and linalyl acetate under normal conditions and after caffeine treatment.

possible with the same finger. In this way, the two main components of response time could be measured separately: "pure" reaction or decision time (RT, time from stimulus onset to release of the first button), and motor time (MT, time from release of the first button to pressing the second button). In the warning condition, a warning tone preceded the visual stimulus.

Ten subjects (5 female and 5 male students between 19 and 25 years old) took part in the experiment, which consisted of 6 sessions in 6 days. In each session, 300 stimuli without warning tone and 300 stimuli with warning tone were presented.   On the 4th and 6th day, the subjects performed their tasks while inhaling the fragrance. Sessions 1 to 3 served as a training procedure establishing a stable level of efficieny.

The results are summarized in Figure 2.   Reaction times with warning (RTW) are always shorter than reaction times without warning tone (RT) because of the increased alertness in the warning condition.   The increase in reaction times from day 3 to day 4 (lavender), the decrease from day 4 to 5 (no lavender), and the increase from day 5 to 6 (lavender) are statistically significant. For the reaction times with warning tone, only the decreases from day 2 to 3 and from day 4 to 5 are significant because of the learning process in the first sessions and the counter-effect of the warning tone against the sedating effects of the fragrance during the following sessions.   Motor times (MT and MTW) do not differ across tests and sessions, indicating that the effect of inhaling lavender is not a peripheral but a central one, influencing only decision times.   Together with the results of the animal experiments discussed above, these data clearly prove the sedative properties of the essential oil of lavender.

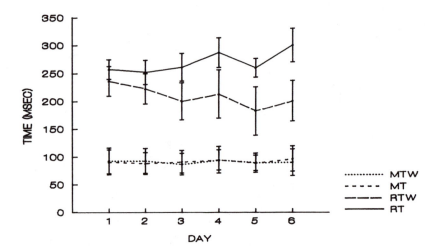

Figure 2.   Means and standard deviations of reaction and motor times. The fragrance (essential oil of lavender) was inhaled on days 4 and 6. MT/MTW - motor times without/with warning tone. RT/RTW -reaction times without/with warning tone.

In addition to theses vigilance experiments, the effect of odors on cerebral bloodflow (CBF) in the limbic system was measured with the stable Xenon-inhalation computerized tomography method (XeCT) (*16*). A concentration of 30% Xe in the inhalation air and a Somatom HiQ-CT (Siemens) were used. Lavender fragrance served as the odor sample. The odor was injected into the airway system after 2 minutes of the xenon flooding phase. The aim of this odor application was to activate the parts of the limbic system involved with olfaction to provoke, as an indirect parameter, a local CBF activation in the corresponding regions in the CT-scan. Until now, no increase of the Local CBF was observed. In order to assess whether the method of XeCT is able to serve for investigation of aromatherapy at all, or whether these first trials support theories maintaining no reflectorial but pharmacological effects of fragrances on the nervous system, further examinations are needed. However, at least we can say that the absence of any change in the CBF (no parts of the limbic system have been activated) is another indication of a direct influence of the fragrance compounds on the brain. This is further argument against effects caused only by a pleasant feeling.

Acknowledgements

We are grateful to:
        Dr. med. Christine Plank, Central Laboratory Animal Facilities, University of Innsbruck, Austria, for performing the animal experiments.

Dr. Elizabeth Karamat, Clinic of Neurology, Department of Psychodiagnosis, University of Innsbruck, Austria; and Miss cand.pharm. Katharina Rösslhuber, Institute of Pharmaceutical Chemistry, University of Vienna, for performing and interpreting the vigilance experiments.

Dr. med. Christian J. O. Nasel, Ludwig Boltzmann Institute of Cerebrovasculary Circulation Research, Vienna; and Primarius Dr. Peter Samec, Department of Radiology, Neurological Hospital Rosenhügel, Vienna, for performing the Xenon Computer Tomography tests.

Austrian Fonds zur Förderung der Wissenschaftlichen Forschung, Project P 8299-CHE, for financial support.

Dragoco Company, Vienna, and Madaus Company, Vienna, both for their financial support and for their encouragement and interest in our work.

Literature Cited

1. Buchbauer, G. *Proceedings of the IFEAT Conference on Essential Oils, Flavors and Fragrances*, Beijing, October 1988, Mapledon Press Ltd., **1989**, pp 350-355.
2. Buchbauer, G. *Perf. & Flav.*, **1990**, *15*, 47-50; and further references cited therein.
3. Buchbauer, G.; Hafner, M. *Pharm. in u. Zeit.*, **1985**, *14*, 8-18.
4. Isacoff, H. *Cosmetics & Toiletr.*, **1981**, *96*, 69-76.
5. Kabara, J. J. *Cosmet. Science, Technol. Ser.* **1984**, *1*, 237-273.
6. Schilcher, H. *Therapiewoche*, **1986**, *36*, 1100-1112.
7. Farag, R. S.; Daw, Z. Y.; Hewedi, F. M.; El-Baroty, G. S. A. *J. Food Protect.* **1989**, *52*, 665-667.
8. Schiffman, S. *10th Internat. Congr. Ess. Oils, Fragrances and Flavours*, Washington, November 16-20, **1986**, general session, pp 47-50..
9. Hanisch, E. *The Nose, part 2*, Firmenschrift Drom, January **1982**, pp 11-18.
10. Gattefossé, R. M. *Parf. Moderne*, **1936**, pp 511-529.
11. *The Concise Oxford Dictionary*, Oxford University Press, **1978**.
12. Wohlfart, R. *Dtsch. Apoth.-Ztg.*, **1983**, *123*, 1637-1638.
13. Kovar, K. A.; Gropper, B.; Friess, D.; Ammon, H. P. T. *Planta Med.* **1987**, *53*, 315-318.
14. Buchbauer, G.; Jirovetz, L.; Jäger, W.; Dietrich, H.; Plank, Ch.; Karamat, E. *Z. Naturforsch.* **1991**, *46c*, 1067-1072.
15. Torii, S.; Fukuda, H.; Kanemoto, H.; Miyauchi, R.; Hamauzu, Y.; Kawasaki, M. In *Perfumery: The Psychology and Biology of fragrance*, van Toller, St.; Dodd, G. H., Eds. Chapman & Hall, London-New York, **1988**, pp 107-120.
16. Drayer, B. P.; Wolfson, S. K.; Reinmuth, O. M.; Dujovny, M.; Boehnke, M; Cook, E. E. *Stroke*, **1978**, *9*, 123-130.

RECEIVED September 9, 1992

FLOWERS

Chapter 13

Volatile Compounds from Flowers

Analytical and Olfactory Aspects

H. Surburg, M. Guentert, and H. Harder

Haarmann and Reimer GmbH, D–3450 Holzminden, Germany

A short survey of the different procedures for the isolation of volatile constituents from fragrant flowers is presented. It is shown, that the three most common methods (dynamic and vacuum headspace sampling, steam distillation) give concentrates whose composition differs both quantitatively and, in many cases, qualitatively. Various results are used by way of example in an attempt to explain how and in what way the different processes lead to different products. The range of usefulness of each method is discussed. For perfumery applications we prefer to use the vacuum headspace technique, which produces odor concentrates with the best sensory properties. Some of the results of these investigations are presented in detail.

In recent years much of the research carried out in the fragrance industry has been directed towards investigation of natural flower fragrances. The aim was to develop a new generation of fragrance compositions more closly related to nature and, thus, to meet the increasing demands of the consumer for more natural products. Consequently, research results gained importance as marketing instruments. But the search for new fragrance substances and new ideas for creating perfume compositions also played an important role in investigations of natural scents.

These research activities were essentially stimulated by the rapid development of modern analytical equipment, which enables us today to analyze complex mixtures in the submicrogram region and to elucidate the constitution of an unknown compound, even if isolated in microgram quantities.

Especially in the case of flowers it appeared that isolation of volatile constituents for analytical and perfumery purposes, using classical methods like extraction and distillation, yielded in most cases, products which did not reproduce the organoleptic properties of the natural material. Consequently, attempts were made to trap fragrant volatiles of flowers directly by so-called "headspace" procedures. However, these methods produced such small amounts of odor concentrates that analysis of the constituents could be accomplished only with the aid of modern analytical techniques.

As a result of this research several papers have been published in recent years dealing with the investigation of flower volatiles. The authors describe the use of very different headspace methods, leading to different, sometimes inconsistent results (1-11).

Until very recently hardly any comparative studies existed *(10,13)*, which attempted to clarify these contradictions; for example, through the application of the different headspace methods to one and the same kind of flower.

Some years ago, we began to perform detailed studies in this field. In the present paper we will describe the experience that we have gained using the different methods for the isolation of volatile constituents of flowers. We will present various results in an attempt to clarify how and in what way the different headspace processes lead to different products, what their advantages and disadvantages are, what risks they involve and how they have to be assessed for perfumery applications.

Different Headspace Methods for Isolation of Volatile Constituents of Flowers

First, a short survey will summarize the methods which are generally used to isolate the volatile constituents of flowers. Since two very detailed reviews of the subject have just been published *(11,12)*, only the principles will be explained briefly.

"Dynamic Headspace". Following dynamic headspace accumulation, the volatiles released by the plant are purged by a gas stream and trapped by adsorption. The adsorption materials used may be charcoal, Tenax or similar macroporous resins. In a second step a concentrate of the volatile compounds is obtained by desorption with a solvent or by heating *(3,5,10,13,14)*.

Very sophisticated variations and modifications of this method have been reported for trapping the fragrant volatiles of living flowers *(5,10,14-18)*. For example, devices have been constructed to enclose a flower which is still connected to the living plant, or the emitted volatiles are sucked off instead of being purged *(19)*.

"Vacuum Headspace". However, since in the products of dynamic processes fewer volatile compounds are found only in reduced quantities or sometimes not at all, a further method was developed: the vacuum headspace method. This method is basically a form of vacuum steam distillation. The flowers are subjected to a vacuum; the volatile components distill off together with the water contained in the plant and are condensed at low temperatures, thus providing the fragrance concentrate immediately *(8,9)*.

Results and Discussion

Comparison of Dynamic and Vacuum Headspace Methods. In recent years we have applied both the dynamic and the vacuum headspace methods for the enrichment of volatile constituents of flowers. To allow comparison we also used in some cases steam distillation at normal pressure, in the modification recommended by Likens and Nickerson. These investigations have demonstrated clearly that not only the quantitative but also the qualitative composition of the respective products differs and depends heavily on the method used. This is shown by the following examples.

Lily of the Valley. The volatile constituents of lily of the valley flowers were enriched in two ways. First we used dynamic headspace sampling during a period of 24 h trapping the volatiles on Tenax and following this by desorption with ether. The second method was the vacuum headspace procedure *(8)*. The composition of the products obtained is shown in Table I which presents some characteristic compounds.

It is evident, that the vacuum headspace concentrate contains higher proportions of higher-boiling, more polar compounds, whereas the low-boiling, less polar components are better represented in the concentrate of the dynamic process. When tested on a smelling blotter, the vacuum product displayed a more typical lily of the valley odor. This may be due to the higher content of compounds like dihydrofarnesol and farnesol.

Hoya carnosa (Wax-Plant). The flowers of *Hoya carnosa* emit their typical odor only at night. To collect the fragrant volatiles, we placed an entire potted plant in a 200 l plastic bag. During the period from 9 p.m. to 6 a.m. a stream of air was passed through this system and led through a Tenax filter at the outlet. In another experiment a flower was cut at midnight and submitted to the vacuum headspace procedure. The composition of the two products obtained is depicted in Table II.

It can be seen that the dynamic process led to higher concentrations of low-boiling compounds like sabinene and linalool. In contrast, the vacuum headspace concentrate contained higher proportions of high-boiling and more polar constituents, e.g. 2-phenylethanol, benzyl alcohol, methyl salicylate, eugenol. The higher content of benzaldehyde in the dynamically obtained product is probably due to an oxidation reaction during the sampling period, because the vacuum headspace concentrate contains smaller amounts of benzaldehyde but larger quantities of benzylalcohol. Again, the product of the vacuum process had superior organoleptic properties.

Phlox. The volatiles of pink Phlox flowers were isolated under the same conditions as described above for lily of the valley flowers. Again, as Table III shows, the dynamic process yields higher proportions of the aldehydic components, whereas the concentrate obtained by vacuum contained higher quantities of the corresponding alcohols. The reason might be an oxidation reaction and/or the higher volatility of the aldehydes in comparison to the alcohols. Cinnamic alcohol could not be detected by the dynamic method; in contrast, the vacuum headspace method produced it in substantial amounts. Curiously, the flowers of white *Phlox* did not contain this substance at all, although the composition is otherwise very similar.

The two isomeric hydroxyketones and the corresponding diol are present only in the vacuum headspace concentrate. In our experience, they are frequently found when this method is used for isolating flower volatiles. As the origin of these compounds is not yet clear, they may be formed as a result of the destruction of the plant material (see below). For the sake of completeness it should be mentioned that lilac alcohols, eugenol methyl ether, isoeugenol methyl ether and germacrene-D are additional main constituents of the volatiles of *Phlox* flowers.

Lilac *(Syringa vulgaris* hybrids). Investigation of the fragrance of purple lilac under the conditions described above produced analogous results, see Table IV. The concentrate obtained by the dynamic process consists mainly of low-boiling compounds like (E)-ocimene and benzyl methyl ether. In contrast, the quantity of these compounds in the vacuum headspace concentrate is much lower. There, the emphasis is on the higher-boiling compounds like lilac aldehydes and alcohols, anisaldehyde etc. Again, the dynamic method produces no cinnamic alcohol; indole and 8-oxolinalool are also missing.

The qualitative composition of the vacuum headspace concentrates from lilac flowers of different colors are generally comparable. Quantitatively, few exceptions can be noted. Red lilac appears to contain only small amounts of ocimene and white lilac produces higher proportions of indole.

Accumulation of Low-Boiling Compounds During Dynamic Processes; Reproducibility of Headspace Methods. The aforementioned results indicate that one of the main differences between the two methods is the higher content of low-boiling compounds in the dynamically obtained headspace concentrate. In order to examine whether and in what way low boiling compounds accumulate during dynamic processes, we investigated the behavior of test mixtures.

The test mixture shown in Table V consists of nearly equal amounts (by weight) of compounds with different boiling points and polarities, as they typically occur in natural flower fragrances. For the experiments a small amount was dabbed on a filter paper. The evaporating substances were purged by a constant stream of air

Table I.   Composition of Headspace Concentrates Obtained by Different Methods from Lily of the Valley Flowers (*Convallaria majalis* L.), [%]

Compound	dynamic	vacuum
(E)-Ocimene	2	<0.05
Citronellyl acetate	18	0.5-3
(Z)-3-Hexenyl acetate	3	~1
Benzyl acetate	5	~1
Benzyl alcohol	10	27-44
Phenylacetonitrile	0.02	2-4
Phenylacetaldoxime	<0.01	1-2
Farnesol	0.2	2-3
2,3-Dihydrofarnesol	0.2	~2
Geraniol	1	10-17
Cinnamic alcohol	0.1	3-7

Table II. Composition of Headspace Concentrates Obtained by Different Methods from *Hoya carnosa* flowers, [%]

Compound	dynamic[1]	vacuum[2]
Sabinene	6	0.6
Benzaldehyde	5.5	1.5
Benzyl alcohol	0.4	14
Isoamyl alcohol	11	25
Linalool	54	23
2-Phenylethanol	trace	4
Methyl salicylate	0.4	2.5
endo-2-Hydroxycineole	0.2	4
Eugenol	-	0.3

1. Living plant, 9.00 p.m. to 6.00 a.m.; 2. Midnight

Table III.   Composition of Headspace Concentrates Obtained by Different Methods from Flowers of *Phlox paniculata* hybrids, [%]

Compound	dynamic[1]	vacuum[1]	vacuum[2]
2-Phenylethanol	0.4	17	12
Phenylacetaldehyde	2	1	1.5
Linalool	2	9	3
Benzaldehyde	5	1	0.3
Benzyl alcohol	0.2	2	0.4
Cinnamic alcohol	-	6	-
3-Hydroxy-4-phenyl-2-butanone	-	3	7
3-Hydroxy-1-phenyl-2-butanone	-	1	4
1-Phenyl-2,3-butanediol	-	0.3	6

1. Pink flowers 2. White flowers

(15 ml/min) and trapped by adsorption on Tenax. This operation was performed repeatedly, whereby the sampling time varied from 20 minutes to 72 hours. As Table V indicates, the compounds of relatively high volatility like linalool and methyl benzoate accumulate after a short sampling time. After a longer period this effect is somewhat compensated. But compounds of a lower volatility and/or higher polarity do not reach the initial values of the test mixture. As a flower continuously produces volatiles, including low boiling compounds, the dynamic process may result in an over-proportional accumulation of compounds of high volatility ("dynamic effect"). Correspondingly, high-boiling compounds are found only in relatively low concentrations. This effect may also be increased by stronger adsorption of high-boiling compounds on the walls of the apparatus.

For comparison we performed a vacuum steam distillation with the test mixture using the tenfold amount of water. The result is shown in the last column of Table V. The data recorded are closer to those of the test mixture; obviously, high-boiling compounds are better represented.

However, when evaluating the results of such model experiments it should be borne in mind that emission of volatiles by a flower is probably not comparable to simple evaporation from the surface of a filter paper. Nevertheless, we believe that our experiments using test mixtures demonstrate the tendency of dynamic processes to accumulate low boiling compounds. It also became clear to us how difficult it is to quantify and thus to compare the results of dynamic processes because they are heavily dependent on the sampling time and/or the velocity of air flow.

With regard to the reproducibility of the vacuum headspace method, we obtained reproducible results without exception, as for example our investigations of the flowers of the damask rose "Rose de Resht" reveal (see Table VI). We observed similar concentrations of the constituents over the whole boiling range. Only the data for geraniol and geranyl acetate differ slightly, but this may be due to the different time of year. It should be mentioned that we picked the rose flowers, like all the roses we investigated, between 1 and 2 p.m. on a warm, but cloudy and not too hot or sunny day. Flowers which had opened in the morning or during the preceding day, gave products with the most beautiful scents.

Artifact Formation.

Dynamic Headspace Method. Our investigation of the volatile constituents of daffodil flowers showed again the familiar picture (see Table VII): The dynamic method produces higher concentrations of low-boiling constituents, whereas the concentrate obtained by vacuum contains correspondingly more high-boiling compounds. A characteristic difference is the occurrence of 2,6-dimethyl-3,5,7-octatrien-2-ol and cosmene among the products of the dynamic process. As we knew ocimene to be an unstable compound sensitive to oxygen, we tested its stability under the conditions of a dynamic headspace process. We found that ocimene is largely oxidized, especially when charcoal is used as the adsorbing material. The main constituents produced by this oxidation reaction are 2,6-dimethyl-3,5,7-octatrien-2-ol and cosmene. In a very recent publication it was reported (20) that even on Tenax monoterpene hydrocarbons like limonene, myrcene and 3-carene suffer considerably from oxidative decomposition and partially disappear. Therefore, the ocimene-releated compounds shown in Table VII may not be of biogenetic origin, but may be artifacts produced by oxidation of ocimene catalyzed by an active surface.

Vacuum Headspace Concentration Compared to Steam Distillation. During our investigations of rose odors, we observed that the composition of the vacuum headspace concentrates are very similar to those of the essential oils obtained by steam distillation with the Likens-Nickerson-method. For example, there is no great difference between the essential oil and the vacuum headspace concentrate of the

Table IV. Compositions of Headspace Concentrates Obtained by Different Methods from Lilac Flowers (*Syringa vulgaris* hybrids), [%]

Compound	dynamic[1]	vacuum[1]	2	3
(E)-Ocimene	85	20-26	17-26	2
Benzyl methyl ether	2	0.2-0.3	0.2	0.1-1
Lilac aldehydes	0.5	8-14	7-8	6-19
Lilac alcohols	0.1	10-17	20-23	36-40
1,4-Dimethoxybenzene	7	6-9	5-9	4-8
Anisaldehyde	0.02	0.6-3	0.3	tr-0.4
Cinnamic alcohol	-	1	1-4	3
Elemicin	0.02	0.2-1	1-6	0.1-1
8-Oxolinalool	-	0.6-3	1-2	1-4
Indole	-	1-2	4	1-2

1. Purple, 2. white, 3. red flowers

Table V. Composition of Headspace Concentrates Obtained by the Dynamic Method from a Test Mixture Depending on the Sampling Time [%]

Compound	0	0.5	6	24	72	[h]	VacHS
Benzyl alcohol	9	10	12	16	15		12
2-Phenylethanol	8	4	7	11	13		10
Linalool	7	22	17	17	12		11
Methyl benzoate	7	44	36	19	13		12
Benzyl acetate	7	13	16	16	12		10
α-Terpineol	7	4	6	9	10		9
Cinnamic alcohol	8	0.5	0.5	0.5	2		4
Indole	9	0.5	1	3	5		6
Geraniol	8	1	1	3	6		8
Geranyl acetate	7	1	2	3	5		6
Methyl cinnamate	6	1	1	2	3		5
Nerolidol	3	-	-	0.2	0.3		1
Benzyl benzoate	8	-	-	0.1	0.1		1
Eugenol	7	-	0.5	1	3		5

Table VI. Volatile Constituents of the Flowers of *Rosa damascena* "Rose de Resht" Obtained by Vacuum Headspace Concentration, [%]

Compound	1	2
2-Phenylethanol	37.4	38.9
Geraniol	35.6	29.8
Citronellol	1.1	0.71
β-Elemene	0.12	0.19
Elemol	0.95	1.25
α-Eudesmol	0.20	0.21
β-Eudesmol	0.13	0.14
Geranyl acetate	0.94	0.59
(Z)-8-Heptadecene	0.51	0.40
Heptadecane	0.59	0.73
(Z)-9-Nonadecene	1.3	1.3
Nonadecane	1.5	1.6

1. June 13, 1989; 2. Sept. 19, 1991

rose cultivar "Fragrant Cloud" as shown in Table VIII. However, two exceptions should be noted. First, β-damascenone was detected in the essential oil but not in the vacuum product. This result agrees well with earlier published observations (21), according to which β-damascenone should be produced from a corresponding precursor under the drastic conditions of steam distillation and is therefore found in the essential oil in higher quantities than in the extract.

The other remarkable difference is the absence of rose oxide in the vacuum concentrate. This compound was also absent in all the concentrates that we obtained by that method from various types of rose, to date more than 30. Other authors (4,10,11) reported that rose volatiles trapped by the dynamic process contain rose oxide as well as some other compounds that we have also not yet detected, e.g. theaspirane, rosefuran, perillene etc. The reason for this discrepancy is still unclear. However, since oxidative reactions may occur during dynamic headspace sampling (see above), it should be checked whether these compounds are really genuine or artifacts.

To allow comparison, the last column of Table VIII shows data reported in the literature (14) for a dynamically obtained odor concentrate. The relatively low content of 2-phenylethanol and the comparatively high values of 2-phenylethyl acetate and 3-hexenyl acetate are striking. This result may be due to the dynamic effect or to the time of sample collection. Investigations by Kaiser (11) have shown that the "Fragrant Cloud" rose produces 3-hexenyl acetate preferentially in the morning.

The influence of heat during steam distillation can also be exemplified by our results with the isolation of the volatile constituents of marjoram leaves by the vacuum headspace method. The data of Table IX show that the content of sensitive compounds like cis-sabinene hydrate and the corresponding acetate is much lower in the steam-distilled essential oil than in the vacuum headspace concencentrate. Instead, higher concentrations of terpinenol-4 are found. It is well known that the latter compound may be formed by rearrangement of sabinene hydrate under slightly acidic conditions for example. Unlike the vacuum headspace concentrate, the essential oil also contains a reasonable quantity of p-cymene, which is sometimes a typical artifact produced by steam distillation. Consequently, the organoleptic properties of the vacuum headspace concentrate are closer to those of the fresh herb. It should be noted that the compositions are comparable to a carbon dioxide extract both qualitatively and quantitatively (see second column of Table IX).

Artifacts Produced by the Vacuum Headspace Method . A major disadvantage of the vacuum headspace method is the fact that the plant material is destroyed. As a consequence of this process, enzymatic reactions could be started which lead to the formation of substances not emitted by the living flower. Such events are illustrated by the following examples.

Anemone nemorosa L. (Ranunculaceae, European wood anemone) is a small anemone found mostly in open woods. The little white to yellow flowers appear in the early spring. Some fragrant populations exhibit a typical and strong odor reminiscent of linalool. When such flowers were subjected to the vacuum headspace process, a concentrate was obtained which contained linalool (25%) and corresponding derivatives (30%) as its main constituents. Additionally, the lactone protoanemonin [5-methylene-2(5H)-furanone] was identified in a remarkable concentration (16%). This compound is known to have strongly lachrymatory properties and is therefore undoubtedly not a genuine constituent of the flower odor. It is formed by an enzymatic reaction from the glycoside ranunculin [5-β-D-glucopyranosyloxymethyl-2-(5H)-furanone], which is a compound commonly found in *Ranunculaceae*.

Many plants with fragrant flowers belong to the plant family *Brassicaceae* (mustard family). The isolation of volatile constituents from such raw materials by the vacuum headspace method yields in many cases products which contain mustard oil compounds to some degree. These are liberated from glucosinolates (mustard oil glycosides) by enzymatic reactions which start when the plant material is destroyed.

Table VII. Compositions of Headspace Concentrates Obtained by Different Methods from Daffodil Flowers *(Narcissus pseudonarcissus* L.), [%]

Compound	dynamic	vacuum
(Z)-Ocimene	4	0.4
(E)-Ocimene	75	30
(E)-6,7-Ocimeneoxide	0.2	-
2,6-Dimethyl-3,5,7-octarien-2-ol	0.4	-
Cosmene (Dehydroocimene)	0.4	-
1,4-Dimethoxybenzene	6	5
3,5-Dimethoxytoluene	1	7
(E,E)-α-Farnesene	2	10

Table VIII. Composition of Odor Concentrates Obtained by Different Methods from the Rose Cultivar "Fragrant Cloud" (M. Tantau, 1963), [%]

Compound	vacuum	oil[1]	dynamic[2]
(Z)-3-Hexenyl acetate	0.4	0.2	10
2-Phenylethanol	45	41	15
Citronellol	5	5	5
Citronellyl acetate	0.4	0.8	5
2-Phenylethyl acetate	4	3	20
Geranyl acetate	2	4	?
Geraniol	16	1	?
Rose oxide	-	0.1	?
β-Damascenone	-	0.01	?

1. Likens-Nickerson method 2. (*14*).

Table IX. Composition of Marjoram Oil Obtained by Different Methods from Marjoram Leaves, [%]

Compound	vacuum	CO_2 extract [1]	oil [1]
trans-Sabinene hydrate	4.2	6.32	3.8
cis-Sabinene hydrate	35	40	11
cis-Sabinene hydrate acetate	4	18	1
Terpineol-4	5	9	21
p-Cymene	-	0.4	3
Caryophyllene	3.5	3.5	2.5
Bicyclogermacrene	1.2	2.2	1.4

1. Commercial grades

Matthiola incana (L.) R. Br. (brampton stock, gilliflower) is a popular garden plant originating from the mediterranean region. Its flowers emit a pleasant floral, clove-like odor. However, in the corresponding vacuum headspace concentrate remarkable amounts of nitriles [5-(methylthio)-pentanenitrile, 0.1%; 5-(methylthio)-4- penteneni-trile, 0,8%] and isothiocyanates [1-isothiocyanato-4-(methylthio)-butane, 0.1%; 4-isothiocyanato-1-(methylthio)-1-butene, 0.4%] are detectable. These compounds are well known as the pungent principles of radish. The true odor of the flowers, how-ever, may be attributed to constituents like 2-phenylethanol, anol, eugenol, methyl eugenol, isoeugenol and related compounds.

When enriched by the vacuum method, the volatile constituents of *Cheiranthus cheiri* L. (wallflower) contain the mustard oil compounds 4-(methylthio)-butanenitrile and 1-isothiocyanato-3-(methylthio)-propane in considerable amounts (3 and 6% respectively) in addition to the main constituents benzyl alcohol (49%) and 2-phenylethanol (20%). Isolation of the volatile constituents by Lickens-Nickerson distillation even produces the mustard oils as main constituents (9 and 81% respectively). This illustrates the influence of the increased temperature, which leads to more pronounced artifact formation.

In some cases, artifact formation by enzymatic cleavage reactions is a desired ef-fect, as we found during our investigation of freshly cut, flowering woodruff herb [*Galium odoratum* (L.) Scop. (*Rubiaceae*)]. Using the vacuum headspace method, we obtained a concentrate of volatile compounds with excellent sensory properties. The main constituent was coumarin in a concentration of more than 80%. Coumarin does not occur originally in the plant, but is formed by enzymatic cleavage of the glycosi-de of 2-hydroxycinnamic acid. Again, this process is initiated by the destruction of the plant material. Other volatile constituents of fresh woodruff herb include linalool, borneol and caryophyllene as well as (Z)-3-hexenol and its acetate.

Formation of Artifacts from Reactive Compounds during Extraction and Di-stillation. In the introduction we mentioned that in the case of flowers classic methods like extraction and distillation often yield products with unsatisfactory orga-noleptic properties because their odor differs significantly from that of the natural raw material. This finding may be illustrated by the following two examples which show how reactive compounds which occur originally in the flowers change during extraction and/or distillation. Thus, the sensory impression is influenced by the dis-appearance of these components and/or by the formation of new compounds.

An extract that we prepared from yellow-bell flowers [*Tecoma stans* (L.) H.B.K., *Bignoniaceae*] contained among its volatile constituents nerolidol as the main com-ponent and, additionally, large amounts of neroli aldehyde and 4-hydroxylinalool. The latter compound is frequently detected in vacuum headspace concentrates of flowers. For example, we found it in the flowers of wisteria [(*Wisteria sinensis* (Sims)Sweet, in literature (*8*) 4-hydroxylinalool is reported as an unknown wisteria flower constituent characterized by its mass spectrum], linden tree (*Tilia cordata* Mill.), mountain ash (*Sorbus aucuparia* L.), wayfaring-tree (Viburnum lantana L.), *Viburnum rhytidophyllum* Hemsl. and grape hyacinth (*Muscari armeniacum* Leichtl. ex Bak.).

The co-occurrence of 4-hydroxylinalool and neroli aldehyde in our extract gave us the idea that neroli aldehyde might not be an irregular terpenoid, but might be formed by an intramolecular pinacol-pinacolone rearrangement from 4-hydroxylinalool or a corresponding precursor (see Figure 1). But this is at present just a hypothesis, which has to be verified experimentally. However if this assumption is true, it would mean that neroli aldehyde, which was first isolated from the essential oil of orange flowers and has been used as a fragrance ingredient for some time, does not occur originally in the flowers, e.g. orange flowers, but is formed only under the drastic conditions during steam distillation.

The second example again concerns orange flowers. While analyzing a commercial but largely unadulterated orange flower absolute, we identified a number of nitrogen-containing heterocycles, e.g. 3-phenylquinoline and various alkyl-substituted 3-phenylpyridines (see Table X).

The propyl phenylpyridines were already characterized as trace constituents of peppermint oil (*22*). But the precursors from which these compounds arose remained unclear. In orange flower extract, the heterocyclic components are probably formed by the following pathways (see Figure 2). Friedlaender and Gohring (*23*) reported as long ago as 1883 that anthranilaldehyde reacts very easily with phenylacetaldehyde to give 3-phenylquinoline. It is possible that such a reactivity is responsible for the fact that anthranilaldehyde is hardly detectable in products prepared by classical methods. Consequently, it was discovered as a flower constituent only when the headspace methods were developed.

The alkyl-substituted 3-phenylpyridines are formed easily by a condensation reaction between ammonia, phenylacetaldehyde, and an α,β-unsaturated aldehyde or ketone, as we have verified in model reactions. For example, 2-hexenal, ammonium acetate and phenylacetaldehyde are converted to 3-phenyl-4-propylpyridine and 5-phenyl-2-propyl-pyridine as main products in about equal amounts. The reaction also works with any other α,β-unsaturated aldehyde or ketone.

It is understandable that, in the case of orange flowers, classically processed products differ from the natural original since organoleptically important compounds like anthranilaldehyde, phenylacetaldehyde, and 2-hexenal disappear and pyridine compounds are formed instead.

Advantage of the Vacuum Headspace Method: the Organoleptic Quality. When our perfumers evaluated the organoleptic properties of the odor concentrates using a smelling blotter, they generally preferred the products of the vacuum process because the odor was closer to nature, except for those products which contain larger amounts of artifacts, as described above for some *Brassicaceae*. The concentrates produced by the dynamic method appeared comparatively flat, atypical and less expressive as a result of the accumulation of low-boiling constituents and an occasional lack of some components of low volatility. Obviously, a higher content of less volatile compounds improves the organoleptic quality of the products.

For an explanation of this phenomenon it has to be realized that the emission of volatiles by a flower and the evaporation from a smelling blotter are based on two entirely different mechanisms. The flower emits its volatile substances with the aid of water or steam, possibly as a kind of aerosol. Therefore, it is able to produce a relatively high concentration in the surroundings, even of compounds of normally low volatility. However, it is very difficult to obtain similar concentrations by simple evaporation from a smelling blotter. This can only be achieved by substantially increasing the proportion of such high-boiling compounds in a mixture of volatile substances. Thus, it becomes clear why the odor of a vacuum headspace concentrate, which contains a higher, perhaps over-proportional amount of less volatile constituents, is closer to the natural fragrance. Hence it follows that for a correct evaluation of a dynamically produced headspace concentrate, a smelling blotter is the wrong test instrument. The organoleptic properties of such a concentrate should be checked by spraying it into the air as an aerosol.

The different mechanisms of evaporation also have implications for the compounding of flower fragrances. The more the typical odor of a flower is determined by high-boiling compounds, the smaller the chance of imitating it using a smelling blotter as an evaporation surface, provided, of course, that only those components are used that have been detected in a headspace concentrate. Lily of the valley is, in our opinion, a good example of this type of extreme situation. The typical odor-determining substances of this flower are farnesol and 2,3-dihydrofarnesol. However,

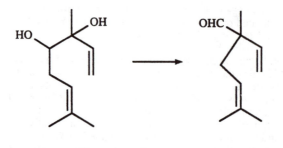

4-Hydroxylinalool   Neroli aldehyde

Figure 1. Possible formation of neroli aldehyde from 4-hydroxylinalool via a pinacol-pinacolone rearrangement.

Table X. Heterocyclic Compounds Detected in Orange Flower Absolute

Compound	Concentration (%)
3-Phenylquinoline	0.01
4-Methyl-3-phenylpyridine	0.001
4-Ethyl-3-phenylpyridine	0.0002
3-Phenyl-4-propylpyridine	0.0005
2-Methyl-5-phenylpyridine	0.001
2-Ethyl-5-phenylpyridine	0.001
5-Phenyl-2-propylpyridine	0.01

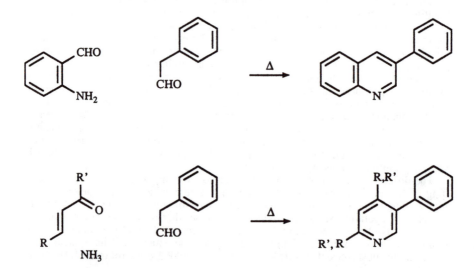

Figure 2. Formation of phenyl-substituted heterocycles from phenyl acetaldehyde.

on a smelling blotter these compounds never achieve the character and the power that they create when sprayed as an aerosol. Yet this is precisely what the flower does.

M. Kerschbaum, the discoverer of farnesol and one of the early fragrance researchers at Haarmann & Reimer, wrote in 1913 when he described this effect, "In the hands of a perfumer farnesol is an obstinate object; using it in compounding perfumes, it is not possible to obtain those effects that nature creates with flowers. The reason might be that the flower emits its fragrance in an infinite dilution, whereas the perfumer following the demand of the public uses the fragrance substances in much higher concentrations." (*24*). We do not believe that this conclusion is correct. On the contrary, the reason is that flowers are able to produce higher concentrations of high-boiling compounds by their special emission technique. The main difference between a flower and a smelling blotter consists in the fact that the flower emits the volatiles all at once, whereas from a blotter compounds evaporate mostly one after the other.

Advantage of Dynamic Headspace Methods: Observation of Changes in the Fragrance Composition. The composition of a fragrance emitted by a flower depends in many cases on the time of day and on the age of a flower. This phenomenon has been thoroughly investigated by several authors (*5,6,11,14,15,17,25*) who used the dynamic headspace method and constructed some sophisticated devices for trapping the volatiles of a living flower. The vacuum headspace procedure is not suitable for such investigations, as the following example illustrates. At 2.00 p.m. we picked a *Hoya carnosa* flower (see above) and submitted it to a vacuum headspace process. Surprisingly, the composition of the concentrate obtained is roughly comparable to that of the fragrant flowers (see Tables and XI), although *Hoya carnosa* flowers are odorless during the day. Obviously, *Hoya carnosa* stores the volatiles produced in the daytime and releases them at night. Since the vacuum process destroys the plant material, all the volatile constituents stored in the plant at that time are caught and not only those, which a flower emits at a definite moment. Therefore, the vacuum headspace method seems to be unsuitable for investigating odor variations during the life of a flower.

Some Applications of the Vacuum Headspace Method. Since our aim was preferably to obtain products with good organoleptic properties as a base for perfumery work, and the vacuum headspace method is a simple procedure giving reproducible results, we employed this process for the investigation of volatile constituents of more than a hundred kinds of flowers and related natural raw materials. In the following, some of the results of these investigations will be presented and briefly discussed.

In the first publication concerning the isolation of flower volatiles using the vacuum headspace method, Joulain (*8*) reported on the identification of a number of uncommon compounds which were not known as flower constituents until that time. This result may be due to the careful isolation conditions, but it is also possible, that some compounds are only released when the plant material is destroyed, e.g. the hydroxylinalools. Therefore, this method is an excellent tool for the isolation of products which are unstable during normal steam distillation, e.g. terpenoid metabolites. The following examples present some rather uncommon compounds that we have detected during our investigations.

The vacuum headspace concentrate of *Reseda odorata* L. flowers contains remarkable amounts of the theaspiranones, shown in Figure 3. These compounds were identified some time ago as constituents of tobacco aroma (*26,27a*). Recently, the second compound was identified as a norisoprenoid aglycon in Riesling wine (*27b*).

Our investigation of the volatile constituents of a broom variety with light yellow flowers (*Cytisus x praecox* Bean) resulted in the identification of N-methyl anthranilaldehyde in a quantity of about 3 % in addition to the main constituents 2-phenylethanol, 2-phenylethyl acetate, anthranilaldehyde, indole, eugenol, and elemicin. Until very recently N-methyl anthranilaldehyde was not known to occur naturally, which is somewhat surprising since the N-methyl derivative of methyl anthranilate has been known for a long time to be a natural compound. It was not until 1989 that Gerlach and Schill (28) reported that they had identified N-methyl anthranilaldehyde among the volatile constituents of a flower of an as yet unclassified orchid species of the genus *Coryanthes*, found in the "Choco" district of Columbia.

Corydalis cava (L. emend. Mill.) Schweigg. et Koerte (*Papaveraceae*) is a perennial herb which grows in the woods of central Europe. Its flowers emit a fine floral-balsamic odor with a slight undertone of amine compounds. In the vacuum headspace concentrate obtained therefrom we identified methyl cinnamate as the main constituent (46%) and in lower concentrations methyl dihydrocinnamate (4%) and methyl phenyllactate (2%). The last is a degradation product of phenylalanine and probably the biogenetic precursor of methyl cinnamate. To date, it has rarely been found in nature. Further main volatile constituents of *Corydalis cava* flowers include 2-phenylethanol, linalool and 2-(4-methoxyphenyl)-ethanol.

Mahonia aquifolium (Pursh) Nutt. (*Berberidaceae*) is an evergreen shrub with little yellow flowers which appear in spring. Their odor is described as sharp amylic, honey-like with chamomile notes. The odor concentrate contained phenylacetaldehyde, 2-phenylethanol, (E)-ocimene and cinnamic alcohol as its main constituents. Additionally, we identified 3,4,5-trimethoxytoluene, a relatively uncommon flower constituent in a remarkable concentration of about 12%. Recently, this compound was detected among the volatile constituents of safflower blossoms (29). Other polymethoxybenzenes are more frequently found during investigations of flower volatiles, e.g. 1,2,4-trimethoxybenzene in hyacinths (1), lilac (3), snowball* (*Viburnum x burkwoodii* Burkw. et Skipw.); 3,5-dimethoxytoluene in yellow roses and related sorts (7,30), narcissus (6), sour cherry blossoms* (*Prunus cerasus* L.) and mimosa*; 1,3,5-trimethoxybenzene in olive tree blossoms (8) and some red roses* (*results of these authors, unpublished work).

Grape hyacinth (*Muscari armeniacum* Leichtl. ex Bak., *Liliaceae*) is a small plant which flowers in spring. Its flowers give off a fruity, floral, slightly peach-like odor. 2-Phenylethanol and 2-(4-methoxyphenyl)-ethanol are the main constituents of the corresponding odor concentrate (see Table XII). Acetophenone and 1-phenylethanol are responsible for the typical topnote. In addition to the well-known 2-phenylethyl benzoate we also identified 1-phenylethyl benzoate, a compound hitherto found only once in nature, namely as a constituent of *Piper hookeri* (31).

Skimmia japonica Thunb. is a evergreen shrub native to Eastern Asia. The little white flowers emit a characteristic floral, lily of the valley-like odor. In keeping with this impression we found farnesol and 2,3-dihydrofarnesol (see Table XIII) exactly as in lily of the valley flowers. Together with nerolidol and 2-phenylethanol these compounds create the typical note. Other main constituents like phenylacetaldoxime and phenylacetonitrile constitute an additional similarity to lily of the valley. 5-Methyl-2-heptanone and the related alcohol are important for the topnote, displaying a fresh, fatty and slightly citrus-like odor. It is remarkable that these compounds have not previously been identified as flower constituents. The first was identified as a trace component in a basil oil (32), the latter as a volatile constituent of roasted beef flavor (33).

The typical odor of German chamomile [*Chamomilla recutita* (L.) Rauschert] flowers is attributed to a number of low-boiling components, which occur in the essential oil in only very small amounts, e.g. ethyl and propyl 2-methylbutyrate. These compounds are found in much higher concentrations in the vacuum headspace concentrate of fresh flowers (see Table XIV). But methyl salicylate, 2-methoxy-4-methylphenol and indole also contribute essentially to the characteristic chamomile note, even though they are present in very low concentrations. While monoterpene hydrocarbons such as the ocimenes and p-cymene appear at comparatively high

Table XI. Composition of a Vacuum Headspace Concentrate Obtained at 2.00 p.m. from a Non-Fragrant *Hoya carnosa* Flower

Compound	%
Sabinene	0.1
Benzaldehyde	0.1
Benzyl alcohol	29
Isoamyl alcohol	35
Linalool	4
2-Phenylethanol	4
Methyl salicylate	1
endo-2-Hydroxycineole	18
Eugenol	0.5

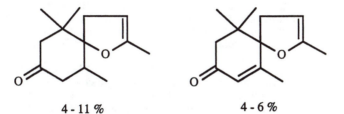

4 - 11 %   4 - 6 %

Figure 3. Theaspiranones identified in the vacuum headspace concentrate of *Reseda odorata* flowers.

Table XII. Compounds Identified in the Vacuum Headspace Concentrate of the Flowers of Grape Hyacinth (*Muscari armeniacum*)

Compound	%
Acetophenone	2
2-Phenylethanol	20
1-Phenylethanol	1
4-Hydroxylinalool	3
2-(4-Methoxyphenyl)-ethanol	31
(E,E)-α-Farnesene	2
1-Phenylethyl benzoate	3
2-Phenylethyl benzoate	2

Table XIII. Compounds Identified in the Vacuum Headspace Concentrate of Flowers of *Skimmia japonica*

Compound	%
Farnesol	19
2,3-Dihydrofarnesol	1
Nerolidol	12
Phenylacetaldehydoxime	13
Phenylacetonitrile	7
2-Phenylethanol	22
Germacrene-D	2
5-Methyl-2-heptanone	2
5-Methyl-2-heptanol	1

concentrations in the headspace concentrate, the content of the higher-boiling main constituents generally corresponds to that of the essential oil, with the exception of chamazulene, a known artifact produced only under steam distillation conditions.

Iceland poppy (*Papaver nudicaule* L.) is a poppy originating from Arctic regions and is now a popular garden plant due to its beautifully colored flowers. Their odor is strong, very chemical, smoky, and clove-like. This sensory impression is reflected by the constituents identified in the odor concentrate (see Table XV), mainly by 2-methoxy- and 4-methoxyphenol, their corresponding methyl ethers and eugenol, which is by far the main component. The topnote is influenced by smaller amounts of acetophenone and 1-phenylethanol.

Dames-violet or sweet rocket (*Hesperis matronalis* L.) belongs to the family of *Brassicaceae*. It is found frequently in old gardens. The color of the flowers which appear in the early summer varies from light purple to lilac. In the evening, the flowers develop a pleasant green-floral, clove-like odor. Accordingly, the following compounds were found in the odor concentrate: benzyl alcohol, 1,8-cineole, benzyl acetate, linalool, α-terpineol, cinnamic alcohol, eugenol, isoeugenol, isoeugenol methyl ether and benzyl benzoate. It should be noted that the concentration of the cis-isomers of isoeugenol and its methyl ether is about 3-5 times higher than that of the trans-isomers (0.5 to 0.1% for the former and 0.2 to 0.07% for the latter). The content of long-chained aliphatic compounds such as hexadecanal (0.2%), hexadecanol (0.2%) and hexadecyl acetate (0.5%) demonstrates again, how well even high-boiling constituents are recovered by the vacuum headspace method.

The flowers of the common cherry laurel (*Prunus laurocerasus* L.) emit a typical cherry odor combined with a pronounced anthranilate note. By far the main constituent of the odor concentrate is benzaldehyde. The impression of anthranilate-like compounds is caused by 2-aminoacetophenone, which we previously detected in the flowers of *Prunus padus* L. (bird cherry) too (*34*). Large amounts of lilac aldehydes and alcohols and linalool derivatives oxidized at C-8 (8-Oxolinalool, (E)- and (Z)-8-hydroxylinalool, 6,7-dihydro-8-hydroxylinalool) are also found. These compounds have also been identified as constituents of the flowers of bird cherry, a related plant. The occurrence of various 4-oxoisophorone derivatives (see Figure 4) proved to be a characteristic property of cherry laurel blossoms. All these compounds are found in a range of 0.1 to 1%.

Erysimum x allionii (*Brassicaceae*) is a plant closely related to the wallflower. The deep orange-colored flowers display a heavy floral odor, but without the anisic violet note of wallflower. The quality of the identified constituents confirmed the organoleptic impression (see Table XVI). It should be mentioned that, in contrast to wallflower, no mustard oil compounds could be detected in the odor concentrate.

Table XIV. Comparison of the Constituents of the Essential Oil and a Vacuum Headspace Concentrate of German Chamomile Flowers [%]

Compound	vacuum	oil [1]
Ethyl 2-methylbutyrate	1.0	0.1
Propyl 2-methylbutyrate	0.5	0.06
2-Methylbutyl propionate	0.2	tr.
(Z)-3-Hexenyl acetate	0.8	tr.
(Z)-3-Hexenyl propionate	0.3	tr.
Artemisiaketone	10	0.44
Artemisiaalkohol	5	tr.
2-Methylbutyl 2-methylbutyrate	0.7	tr.
Methyl salicylate	0.2	-
2-Methoxy-4-methylphenol	0.4	-
Indole	0.04	-
p-Cymene	0.7	0.1
(Z)-Ocimene	0.9	0.2
(E)-Ocimene	4.3	0.8
(E)-beta-Farnesene	23.0	26.2
Germacrene-D	2.2	0.3
Bicyclogermacrene	2.7	1.1
Bisabolol oxide B	4.0	4.2
Bisabolone oxide	3.1	5.1
Bisabolol	0.9	2.0
Bisabolol oxide A	28.2	6.8
Chamazulene	-	1.2
En-In dicycloether	0.6	3.0

1. Commercial grade

Table XV. Compounds Identified in the Vacuum Headspace Concentrate of Iceland Poppy (*Papaver nudicaule*) Flowers

Compound	%
Acetophenone	0.4
1-Phenylethanol	0.1
2-Methoxyphenol	2
1,2-Dimethoxybenzene	1
1,4-Dimethoxybenzene	6
4-Methoxyphenol	8
Eugenol	45
2-(4-Methoxyphenyl)-ethanol	11
Methyl eugenol	2
2-Heptadecanone	12

Table XVI. Compounds Identified in the Vacuum Headspace Concentrate of *Erysimum x allionii* Flowers

Compound	%
Benzyl alcohol	34
2-Phenylethanol	39
Linalool	1
Methyl salicylate	2
Anthranilaldehyde	0.4
5-Hydroxylinalool	1
Indole	3
Methyl 2-methoxybenzoate	2
Caryophyllene	2
Benzyl benzoate	2

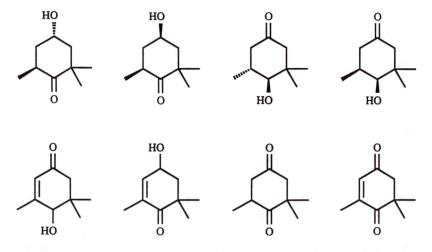

Figure 4. Isophorone derivatives found in the vacuum headspace concentrate of flowers of common cherry laurel

The southern Meditearranean coast of Turkey is home to a narcissus species named "Nergiz" by the locals. It is a small tazetta-like plant with several doubled flowers on each stem. The odor concentrate contained mainly trivial compounds such as benzyl alcohol, (E)-ocimene, benzyl acetate, indole, methyl cinnamate etc. Some uncommon minor components are noteworthy: 2-methoxybenzylalcohol and the corresponding acetate have previously been found only in hyacinth (*1*) and narcissus (*35*) flowers. The homologous 2-(2-methoxy)phenylethyl alcohol is a new natural product. Its acetate is already known, but only as a narcissus constituent (*35*).

Acknowledgement

We would like to thank the management of Haarmann & Reimer for permission to publish this paper, the staff of the H&R research department for their valuable cooperation, especially Miss S. Allerkamp and Mrs. Chr. Voessing for their skilful technical assistance and Dr. P. Werkhoff for helpful discussions. We are grateful to

Dr. P. Calzavara, H&R Italy, for carrying out headspace experiments in the Mediterranean region and Dr. H. Harder and A. Landi for sensory evaluations.

Literature cited

1. Kaiser, R.; Lamparsky, D. *Parfums, Cosmet., Aromes* **1977**, *17*, 71-79.
2. Kaiser, R.; Lamparsky, D. in *VIIIth International Congress of Essential Oils, Cannes/Grasse 1980, Technical Data*; FEDAROM: Grasse, 1982, 287-294.
3. Lamparsky, D. in *Essential Oils and Aromatic Plants*; Baerheim-Svendsen, A. and Scheffer, J.J.C. Eds.; M. Nijhoff/Dr. W. Junk Publishers: Dordrecht 1985; 79-92.
4. Kaiser, R.; *Conference Papers of the International Conference on Essential Oils, Flavours, Fragrances and Cosmetics in Beijing/China, Oct. 1988;* IFEAT: London; 272-289.
5. Mookherjee, B.D.; Trenkle, R.W.; Wilson, R.A.; Zampino, M.; Sands, K.P.; Mussinan, C.J. in *Flavors and Fragrances: A World Perspective;* Lawrence, B.M.; Mookherjee, B.D.; Willis, B.J., Eds., Elsevier Sci. Publ. B.V., Amsterdam, 1988; 415-424.
6. Mookherjee, B.D.; Trenkle, R.W.; Wilson, R.A. *J. Ess. Oil Res.* **1989**, *1*, 85-90.
7. Mookherjee, B.D.; Trenkle, R.W.; Wilson, R.A. *Pure Appl. Chem.* **1990**, *62*, 1357-64.
8. Joulain, D. in *Progress in Essential Oil Research;* Brunke, E.J., Ed.; W. de Gruyter, Berlin, 1986; 57-67.
9. Joulain, D. *Flav. Frag. J.* **1987**, *2*, 149-155.
10. Brunke, E.J.; Hammerschmidt, F.J.; Schmaus, G. *Dragoco report* **1992**, 2-31.
11. Kaiser, R. in *Perfumes, Art, Science and Technology;* Mueller, P.M.; Lamparsky, D., Eds., Elsevier Applied Science, London and New York, 1991, 213-250.
12. Bicchi, C.; Joulain, D. *Flav. Frag. J.* **1990**, *5*, 131-45.
13. Bergstroem, G.; Appelgren, M.; Borg-Karlson, A.K.; Groth, I.; Stroemberg, S.; Stroemberg, St. *Chemica Scripta* **1980**, *16*, 173-180.
14. Flament, I. *Cosmetics and Toiletries manufacture 1991/2*, century press 1991, 114-22.
15. Matile, P.; Altenburger, P. *Planta* **1988**, *174*, 242-247.
16. Loughrin, J.H.; Hamilton-Kemp, T.R.; Andersen, R.A.; Hildebrand, D.F. *J. Agric. Food Chem.* **1990**, *38*, 455-60.
17. Loughrin, J.H.; Hamilton-Kemp, T.R.; Andersen, R.A.; Hildebrand, D.F. *Phytochemistry* **1990**, *29*, 2473-77.
18. Omata, A.; Nakamura, S.; Yomogida, K.; Moriai, K.; Ichikawa, Y.; Watanabe, I. *Agric. Biol. Chem.* **1990**, *54*, 1029-33.
19. Kaiser, R.; Nussbaumer, C. *Helv. Chim. Acta* **1990**, *73*, 133-9.
20. Stroemvall, A.M.; Petersson, G. *J. Chrom.* **1992**, *589*, 395-9.
21. Ohloff, G.; Demole, E. *J. Chrom.* **1987**, *406*, 181-3.
22. Sakurai, K.; Takahashi, K.; Yoshida, T. *Agric. Biol. Chem.* **1983**, *47*, 2307-12.
23. Friedlaender, P.; Gohring, C.F. *Ber. Dtsch. Chem. Ges.* **1883**, *16*, 1833-9.
24. Altenburger, R.; Matile, P. *Planta* **1988**, *174*, 248-52.
25. Kerschbaum, M. *Ber. Dtsch. Chem. Ges.* **1913**, *46*, 1732-7.
26. Demole, E.; Berthet, D. *Helv. Chim. Acta* **1972**, *55*, 1866-82.
27. a. Fujimori, T.; Takagi, Y.; Kato, K. *Agric. Biol. Chem.* **1981**, *45*, 2925-6; b. Winterhalter, P.; Sefton, M.A.; Williams, P.J. *J. Agric. Food Chem.* **1990**, *38*, 1041-8.
28. Gerlach, G.; Schill, R. *Pl. Syst. Ecol.* **1989**, *168*, 159-65.
29. Binder, R.G.; Benson, M.E.; Flath, R.A. *J. Agric. Food Chem.* **1990**, *38*, 1245-8.

30. Nakamura, S. *Perfumer and Flavorist* **1987**, *12* (05/06), 43-5.
31. Singh, Jagdev; Atal. C.K. *Phytochemistry* **1969**, *8*, 2253-4.
32. Ekundayo, O.; Laakso, I.; Oguntimein, B.; Okogun, J.I.; Elujoba, A.A.; Hiltunen, R. *Acta Pharm. Fenn* **1987**, *96*, 101-6.
33. Hsu, Ch. M.; Peterson, R.J.; Jin, Q. Zh.; Ho, Ch. T.; Chang, St. S. *J. Food Sci.* **1982**, *47*, 2068.
34. Surburg, H.; Guentert, M.; Schwarze, B. *J. Ess. Oil Res.* **1990**, *2*, 307-16.
35. Sakai, T. *Shitsuryo Bunseki* **1979**, *27*, 202R-5R *(Chem. Abstr.* **1980**, *93*, 173591).

RECEIVED August 19, 1992

Chapter 14

Cryogenic Vacuum Trapping of Scents from Temperate and Tropical Flowers

Facts and Figures

D. Joulain

Research Laboratories, Robertet S. A., B.P. 100, F–06333 Grasse Cedex, France

In 1985, we introduced a very convenient method for trapping the volatile compounds emitted by fragrant flowers. To date, this method has been applied successfully by us and others to a large number of floral species, originating from temperate as well as from tropical countries. We present now further results obtained from local flowers : jonquil, jasmin, rose, and osmanthus, cultivated in the South of France, and *Gardenia* and *Jasminum* species from tropical countries. The results confirm the value of the method since it allows to detect previously unreported constituents in both the "headspace" or extracts from these flowers. Some quantitative data dealing with model mixtures, using the cryogenic vacuum trapping method, are also discussed.

The fragrance of flowers, in association with their appearance, has always been a subject of fascination for mankind and people have always referred to it in the most poetic way. As soon as it became possible, scientists wanted to know what elements are responsible for the olfactive impression given in a characteristic manner by each fragrant flower species. It is therefore understandable that many of those scientists are members of the perfume industry.

The first research work on the direct analysis of volatile compounds diffused by flowers was undertaken ca. 30 years ago; it followed closely in the footsteps of gas chromatography. The greatest progress has been accomplished over the last 15 years, due to an increasingly efficient use of capillary gas chromatography coupled with mass spectrometry (*1, 2*). More recently, the capability to elucidate chemical structures of substances that are only available in minute quantities within complex mixtures has improved further through other state of the art techniques such as GC/FTIR and automated micropreparative capillary GC. The latter technique gives access to NMR spectrometry with which one can almost always positively determine a structure.

0097–6156/93/0525–0187$06.00/0

Techniques enabling isolation of the total fragrant vapor phase emitted by flowers ("headspace") are diverse (2). In this presentation we shall emphasize the efficiency of the "vacuum cryogenic trapping" method (VCT), also referred to as vacuum degassing and low temperature trapping. A typical experimental procedure has been described (3). Since 1985, we have been using this methodology successfully with over 150 fragrant flowers species. We are now able to confirm that the odor of the trapped scents was always very close, if not identical in many cases, to that of the living flowers, provided the time of the processing had been suitably chosen for each type of flower.

We have examined not only fragrant flowers which are or have been formerly processed in the perfume industry, but also other types of flowers originating from temperate areas such as the Grasse region (southeastern France) as well as from tropical countries.

Jonquil

In Grasse, jonquil flowers (*Narcissus jonquilla* L.) are extracted industrially with hexane . The odor of the crude solvent-free extract ("concrete") is described as deep-sweet floral with a strong green undertone and is reasonably reminiscent of the fresh flowers. Nine compounds alone represent ca. 93% of the headspace isolated by VCT, among which we have recorded high levels of methyl benzoate and indole (Table I). The most volatile fraction is mainly characterized by several contributors to the green note: isoprenol, prenol, 3(Z)-hexen-1-ol (major), prenyl and isoprenyl acetates, methyl 3(Z)-hexenoate etc...

Jasmine

Jasmine and rose flowers remain the two most important floral raw materials used in high class perfumes. For this reason, over the years, they have been subjected to very thorough analytical investigation. In Grasse, flowers of *Jasminum grandiflorum* L. grafted on *J. officinalis* L. are still processed industrially from July to late October.

To date, a detailed analysis of the headspace from jasmine flowers (*J. grandiflorum* L.) has not given rise to any publication. However, while "living flowers" headspace has been reported to contain up to 60% benzyl acetate, 11% indole, 3% jasmone and 0.3% methyl jasmonate among others, that of picked flowers contains less indole and neither jasmone or methyl jasmonate, two important contriutors to the jasmine odor (4).

Significantly no information was provided in this report, however, concerning the hour of the day this experiment was carried out. It is a well-known fact that the frarance given off by jasmine flowers, as well as in the case of many other flowers, is strongly dependent in terms of intensity and quality on the time of day. The fragrance develops during the night, and seems to reach a maximum at the break of day, precisely when harvesting takes place. Starting each time with 50g of flowers from a given growing area, we isolated the headspace (VCT method) in three steps at 6 hours intervals, and we measured the total quantity of trapped volatiles, using an internal standard (methyl nonanoate). Results are shown in Table II.

One can see that the bulked "jasmonoids" (jasmone, jasmin lactone and methyl jasmonate) show a maximum concentration early in the morning, thus confirming what has been known for many years by jasmine producers in Grasse. Of particular interest is the presence of methyl epi-jasmonate **1** (Figure 1), practically as a single isomer (97%, versus 3% of methyl jasmonate). This indicates that methyl jasmonate is very likely an artifact when reported as a component of the extracted jasmine oil, or of the headspace when isolated by methods other than VCT. During these processes, methyl epi-jasmonate appears to be rapidly isomerized into the thermodynamically more stable methyl jasmonate possessing a trans configuration.

Among the other fragrant *Jasminum* species, *J. sambac* (L.) Aiton is of some importance, mainly in Southeast Asia including China, where it is highly praised for its ornamental and fragrant qualities. Several reports dealing with the Dynamic Headspace (DHS) analysis of the fragrance of this flower have appeared in the literature (5-7). Flowers from Thailand, Malaysia and Singapore have been made available to us, and we were able to isolate the headspace in good conditions using VCT. A total of 65 components were readily identified, 14 of which at higher percentage than 1% (Table III). Among a number of previously unreported minor compounds, we can mention 5-oxolinalool (0.03%) and 5-hydroxylinalool (0.1%). These results are significantly different from those reported by the Chinese authors.

Rose

Hybrids of *Rosa centifolia* L. are cultivated in the Grasse area, for industrial processing by hexane extraction. The characteristic odor of the absolute oil is mainly due to 2-phenylethanol, geraniol, nerol and citronellol in high concentrations and to a number of other constituents in smaller proportions. High levels of all these elements are also found in the headspace of the same flowers (Table IV, Figure 2).

Developing rose hybrids that could find applications in the perfume industry, either as new fragrant models for creating perfumes or as new raw materials, is still an attractive goal. Systematic investigations for producing original rosy effects have lead to surprising results, for the headspace of some of newly created hybrids, although possessing an undisputable rose odor, can be totally devoid of the well-known above-mentioned rose odor contributors. As an example, one can mention a new rose, the headspace of which is characterized by 4 constituents only, accounting for ca. 90% of the volatiles (Table V, Figure 3), if not taking in account the odorless hydrocarbons with at least 15 carbon atoms. The odor is described as rosy with anise-like, spicy and green undertones.

Osmanthus

The fragrance of *Osmanthus fragrans* fragrans Lour. (Oleaceae) is exceptionally

Table I. Main constituents of jonquil headspace (VCT)

	%a		%a
(E)-ß-ocimene	35.3	methyl (E)-cinnamate	7.8
methyl benzoate	23.4	prenyl benzoate	1.8
linalool	17.8	(E)(E)-α-farnesene	0.6
benzyl acetate	1.0	benzyl benzoate	4.4
indole	1.7		

a = Relative percentages as recorded by the GC integrator

Table II. Variation of selected J. grandiflorum flowers odorants with time of day

Time	7:00 a.m.	1.00 p.m.	7.00 p.m.
Total volatiles (± 15 ppm)	150	50	90
paracresol (%)	2.3	0.4	6.8
linalool (%)	6.7	25.3	13.4
benzyl acetate (%)	65	51	57
indole (%)	6	3.1	4.3
jasmin lactone (%)	0.8	-	0.07
jasmone (%)	3.2	0.5	0.6
methyl epi-jasmonate (%)	0.2	-	-

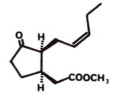

1

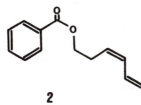

2

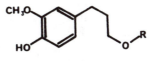

3 *R = H*

4 *R = Ac*

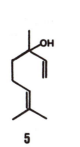

5

6

7

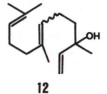

8

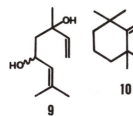

9

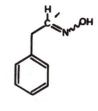

10

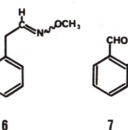

11

12

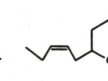

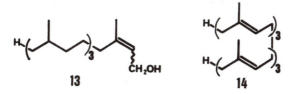

13

14

Figure 1. Structural formulas.

Table III. Main constituents of *J. sambac* headspace

linalool	37.1 %	3(Z)-hexen-1-yl benzoate	2.9%
(E)(E)-α-farnesene	14.2	4,8,12-trimethyltrideca-	
indole	10.7	1,3,7,11-tetraene	2.0
1(10),5-germacra-	8.6	methyl salicylate	1.7
dien-4-ol		3(Z)-hexen-1-yl-acetate	1.5
benzyl acetate	4.9	3(Z)-hexen-1-ol	1.0
para-cymene		germacrene D	1.0
benzyl alcohol	3.3	methyl anthranilate	1.0

Table IV. Main constituents of the headspace of *Rosa centifolia* flowers

1 prenol	16 geraniol (7.8 %)
2 3(Z)-hexen-1-ol	17 geranial
3 α-pinene	18 tridecane
4 myrcene	19 methyl geraniate
5 benzyl alcohol (1.1 %)	20 eugenol
6 1,8-cineole	21 citronellyl acetate
7 limonene	22 neryl acetate
8 (E)-ß-ocimene	23 geranyl acetate (2.6 %)
9 2-phenylethanol (27.3 %)	24 methyleugenol (2.6 %)
10 cis-rose oxide	25 germacrene D
11 4,8-dimethyl-1,3(E),	26 pentadecane
7-nonatriene	27 a heptadecene
12 trans-roseoxide + veratrole	28 heptadecane
13 estragole	29 9(Z)-nonadecene (0.85 %)
14 citronellol + nerol (20.8 %)	30 nonadecane (0.92 %)
15 2-phenylethyl acetate	

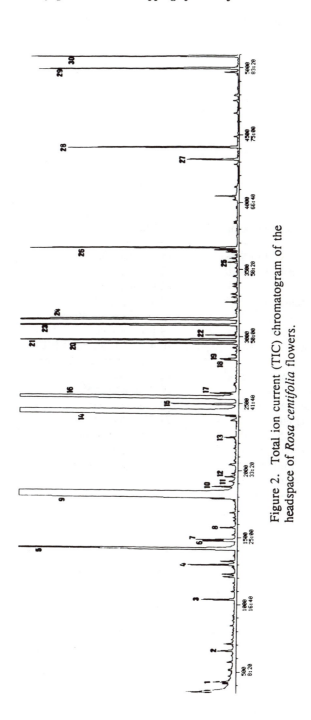

Figure 2.   Total ion current (TIC) chromatogram of the headspace of *Rosa cenifolia* flowers.

Table V. Headspace components from a new rose hybrid

1 3(Z)-hexen-1-ol (2.7 %)	14 α-humulene
2 2(E)-hexen-1-ol	15 ß-ionone
3 1-hexanol	16 germacrene-D
4 3(Z)-hexen-1-yl	17 1-dodecanol
acetate (3.2 %)	18 (E)(E)-α-farnesene
5 1-hexyl acetate	19 pentadecane
6 4,8-dimethyl-1,3(E),	20 4,8,12-trimethyl-1,3(E),
7-nonatriene	7(E),11-tridecatetraene
7 4-vinylanisole (29.6 %)	21 1-tridecanol
8 decanal	22 1-tetradecanol
9 2,3-dihydrobenzofurane	23 a heptadecene
10 3,5-dimethoxy-	24 heptadecane
toluene (42.5 %)	25 1-pentadecanol
11 ß-caryophyllene	26 9(Z)-nonadecene (4.3 %)
12 dihydro-ß-ionone	27 nonadecane (1.4 %)
13 geranylacetone	

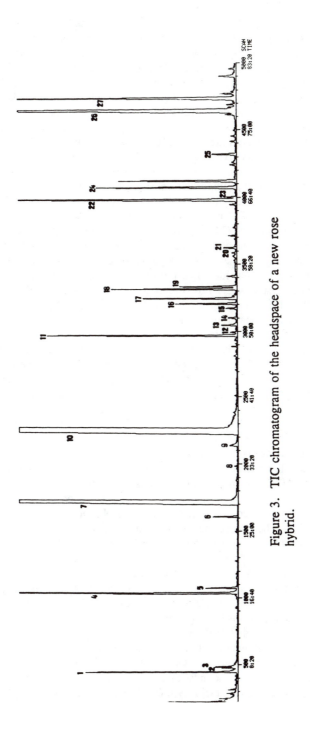

Figure 3. TIC chromatogram of the headspace of a new rose hybrid.

rich and diffusive. The absolute obtained from commercial hexane extracts (mainly of Chinese origin) is a highly valuable raw material, used only in high class perfumes to impart a fruity and warm peach and apricot effect. Different species of osmanthus (a 2-5 m high evergreen tree) can be seen growing freely in gardens in Grasse. Early in the month of November, one species produces golden-yellow flowers possessing a potent ripe apricot fragrance with a tobacco-like undertone. A similar sized silvery-white flower variety gives off a weaker and less diffusive closely related fragrance, with, however, an undertone of a green tea leaf type. These olfactive differences are illustrated by the corresponding chromatograms of the trapped volatile parts. Although gamma-decalactone and ß-ionone, both characteristic contributors to the osmanthus odor, are present at high levels in both flowers, one can see that important constituents are reciprocally present in one species and completely absent in the other. For example, 5-oxolinalool (1.5%) and 5-hydroxylinalool (5.2%), neither ever mentioned before as osmanthus flower components, and geraniol (24.7%) are only present in the orange variety, whereas jasmone (4.0%), ß-ionone photoisomer (7.3%), benzothiazole (0.9%) and nerolidol (2.6%) occur only in the silvery type (Figure 4). One constituent of the silvery-white species, not identified positively by GC/MS alone, appears to be a retro α-ionone. Use of GC/FTIR (Figure 5) confirmed a non-conjugated carbonyl group and the lack of an exocyclic methylene group (Figure 6). The exact stereochemistry however, remains unknown.

The ß-ocimene levels in the orange and silvery types we studied (0.06% and 33% respectively) cannot be taken as discriminating principles, for it is a known fact that this hydrocarbon proportion can vary drastically within a given flower, depending of the hour of the day (2). The above results confirm to a certain extent the previous findings reported with Chinese osmanthus flowers (8).

Gardenia

Within the Rubiaceae family, the *Gardenia* genus produces an abundance of very fragrant white flowers, among which *Gardenia jasminoides* Ellis is certainly the most popular. The first analysis of the headspace of these flowers was reported in 1977 (9). In 1984, we investigated flowers that had been grown in a phytotron (under humid tropical conditions), and we were able to separate and identify 46 constituents in the headspace including mainly : methyl benzoate, 3(Z)-hexen-1-yl benzoate, linalool, benzyl benzoate, jasmin lactone (all previously known) in the order of decreasing proportions.

If is noteworthy that although the processing of the flowers actually took place 5 hours after their picking, methyl benzoate was still the major element (17.6%), while ethyl benzoate was only present in very low proportion (0.2%); this contradicts certain previously published data.

Gardenia tahitensis D.C. is less known, although its flower is very popular as "tiaré" in French Polynesia. Recently, a diethylether extract from these flowers has been analysed, resulting in identification of 27 constituents, including two artifacts and only 7 compounds that can be looked upon as fragrant with 3(Z)-hexen-1-yl benzoate in majority (10). We carried out an in-depth analysis of

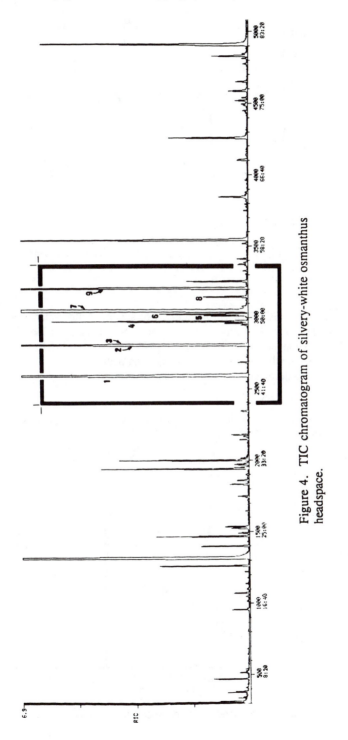

Figure 4. TIC chromatogram of silvery-white osmanthus headspace.

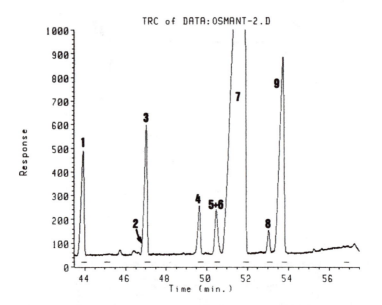

Figure 5.   GC/FTIR chromatogram of silvery-white osmanthus flowers.   No. 1. ß-ionone photoisomer; No. 2.jasmone; No. 3. a retro-α-ionone ?; No. 4. α-ionone; No. 5. 6-pentyl-2-pyrone; No. 6. dihydro-ß-ionone; No. 7. gamma-decalactone; No. 8. δ-decalactone; No. 9. ß-ionone.

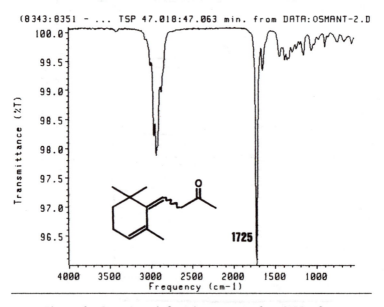

Figure 6.   Gas-phase infrared spectrum of peak No. 3.

the absolute from "tiaré" flowers from Tahiti, prepared by industrial means under our supervision. We were able to separate and identify over 200 constituents forming a group mainly consisting of 2,3-epoxysqualene (major) and various benzoates and salicylates (Joulain, unpublished results). From flowers picked in Polynesia, in the Wallis-and-Futuna islands, Northeast of the Fiji islands, we isolated the volatile part using our standard method, and identified 68 constituents (Table VI, Figure 7).

Apart from high levels of linalool (23%), methyl salicylate (13.5%) and 3(Z)-hexen-1-yl benzoate (21.3%), we recorded the presence of smaller amounts of phenylacetaldoxime-methylether, 2-phenyl-1-nitroethane and phenylacetonitrile (7% together) already mentioned (2). Also noteworthy is the presence of 3(E), 5-hexadienyl benzoate **2** (0.2%), Figure 1, previously reported in these flowers and in *Jasminum sambac* (6), as well as dihydroconiferyl alcohol **3** (trace) and its acetate **4** (0.1%) never mentioned before, and that seem to be characteristic of this gardenia species.

Most of the quantitative data reported in the literature consist of relative chromatographic percentages of eluted substances, which assumes they all possess the same response factor (equal to 1) within a given detection device. Actually, very little work has been undertaken to confirm the accuracy of these measurements and their reproducibility, or to compare the relative merits of the different methods that can be used to collect the volatile compounds from flowers.

Qualitative differences have been observed when treating kiwi flowers (*Actinidia chinensis* Planch.) by steam-distillation/extraction (SDE) or by the dynamic headspace method (DHS). When using SDE, artifacts can occur by disruption and heating of the flower tissues (*11*). Very significant qualitative and quantitative differences have been recently reported when using DHS, SDE and VCT for isolating the volatiles from *Daphne mezereum* L. flowers (*12*). The same phenomenon was observed when treating lilac flowers (*Syringa vulgaris* L.) by either VCT or "closed loop stripping" DHS (*13*).

We have tested the recovery levels of 10 selected substances by SDE and VCT. Compounds 5-14 (Figure 1) occur in many flowers and have been chosen because of their wide range of volatility, and in the case of some of them, for their suspected sensitivity to the experimental conditions (probability of artefact formation). All the following experiments were performed in triplicate.

a.   A 3 hour SDE was conducted at normal pressure starting with ca. 100 mg of mixtures of the compounds to be tested in known quantities and 100 ml water. The extraction solvent used was dichloromethane (analytical grade, redistilled), to which, at the end of the procedure, known quantities of two internal standards, veratrole and methyleugenol, were added. The former was used to determine the amounts of linalool **5**, phenylacetaldoxime-methylether **6**, 2-aminobenzaldehyde **7**, phenylacetaldoxime **8**, and ß-ionone photoisomer **10**, while the latter was used for 5-hydroxylinalool **9**, jasmin lactone **11**, nerolidol **12**, phytol **13** and squalene **14**.

b.   VCT experiments were conducted at 0.5 Torr pressure and 20°C over a 45 mm period, starting with ca. 100 mg of mixtures as above, and 20 g water. The same internal standards were added at the extraction stage. Gas

Table VI. Main constituents of *Gardenia tahitensis* D.C.
headspace (selection)

1. 3(Z)-hexen-1-ol	13. methyl salicylate
2. α-pinene	14. 2-phenyl-1-nitroethane
3. camphene	15. 3(Z)-hexen-1-yl tiglate
4. ß-pinene	16. 3(Z)-hexen-1-yl benzoate
5. 4-methoxytoluene	17. hexyl benzoate
6. 1,8-cineole	18. 3(Z),5-hexadien-1-yl
7. limonene	benzoate
8. methyl benzoate	19. 3(Z)-hexen-1-yl
9. trans-linalyl	salicylate
oxide (furan.)	20. dihydroconiferyl acetate
10. linalool	21. tetradecanoic acid
11. phenylacetonitrile	22. 2-phenylethyl benzoate
12. phenylacetaldoxime-	23. 2-phenylethyl salicylate
methylether	24. hexadenoic acid

chromatographic response factors and concentrations were measured by using a non-polar (SE 30) fused silica capillary column (30 m x 0.32 mm i.d.), and a micro-cell thermal conductivity detector and helium carrier gas. Results are summarized in Figure 8.

One can see that SDE gives fairly acceptable results for products of high and medium volatility only, in spite of the risk of the formation of artefacts for sensitive products. For example, compounds **8** and **9** are partially transformed into phenylacetaldehyde and hotrienol respectively under these conditions. The reliability of the VCT experiment can be questioned as a suitable model, since it is practically impossible to reproduce artificially the complex matrix of living flower tissues. Excellent to good recovery levels of all products are observed however, when using the VCT method, with the exception of phytol **13** and squalene **14** which are, nevertheless, reasonably detected. Blank experiments have furthermore shown that during the second stage of the VCT procedure (dichloromethane extraction, filtration and concentration), not less that 95 % (average) of each of the 10 tested products are recovered.

Similar testing using a DHS procedure proved to be extremely difficult for us to handle, mainly due to the very low quantities of substances involved. When using activated charcoal DHS (adsorption during 3 hrs followed by desorption with carbon disulfide and methanol) and the same test mixtures and internal standards however, we were able to carry out semi-quantitative evaluations. Severe discrimination expectedly occurred between high, medium and low volatility compounds, the latter i.e. phytol and squalene giving insignificant results (Table VII).

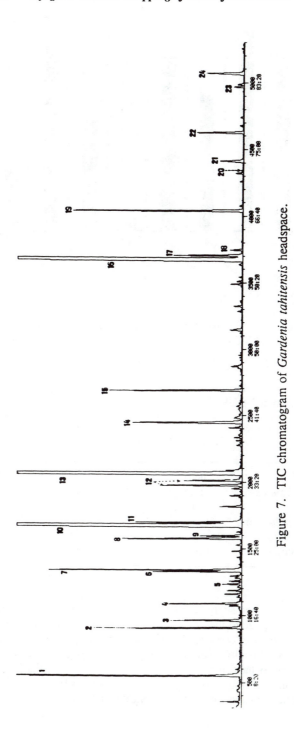

Figure 7. TIC chromatogram of *Gardenia tahitensis* headspace.

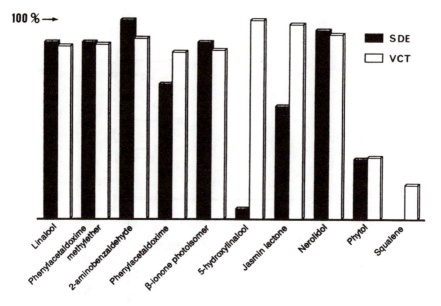

Figure 8. Compared recovery levels of compounds 5-14
SDE and VCT

Table VII. Recovery levels (%) of compounds 5-14 with activated
            charcoal DHS

Good 70-100 %	Fair 40-70 %	Bad < 40 %	Very Bad (no recovery)
linalool phenylacetaldoxime- methylether ß-ionone photoisomer 5-hydroxylinalool jasmin lactone	nerolidol 2-amino- benzaldehyde	phenylacetal- doxime	phytol squalene

 In short, for the study of the fragrance of flowers, SDE is definitely not a suitable technique, due to the risk of artifact generation. Either DHS or VCT can be used satisfactorily to carry out systematic research, provided one given method alone is chosen, to allow comparison of future results. Vacuum Cryogenic Trapping seems to have a decisive advantage since it permits getting a closer approach to the real fragrance of the flower (Table VIII) ; a feature of prime importance when looking for possible applications in perfumes.

Table VIII. Advantages of the VCT method

■Short procedure : ca. 60 mn.
■Good recovery of compounds of medium and low volatility
■Very low risk of artifact formation
■ Compatible with GC/FTIR and micropreparative capillary GC

Table IX. Variations of selected main constituents of Mimosa headspace
according to exposure to sunlight

---

	in the sun (%)a	in the shade (%)a
3(Z)-hexen-1-ol	1.0	2.3
benzaldehyde	6.7	22.5
3(Z)-hexen-1-yl acetate	2.3	29.5
(E)-ß-ocimene	0.8	trace
benzyl alcohol	4.5	0.8
phenylacetaldehyde	n.d.	0.5
linalool	0.7	4.9
anisaldehyde	2.5	2.7
pentadecane	4.2	0.7
heptadecene	77.2	36.1
Total	100.0	100.0

---

n.d. = not detected
a = Percentages are normalized to 100

Thus, when attempting artificial reconstitutions of flower fragrances, many difficulties, mainly quantitative, arise as we have just demonstrated. Most of these problems have been discussed previously (2). It is essential moreover, to define accurately for each flower what model is to be used because a number of aspects have to be taken into account :

1.   Flower type. We have shown above in the case of osmanthus that two varieties can generate different fragrances. The same is true of *Jasminum grandiflorum*, whether grafted jasmine is involved (as in Grasse) or not, or of *J. sambac* that exists as different types (6).

2.   Hour-of-day. Circadian rhythm may have a decisive influence on the composition of the flower volatiles (14), as shown above in the case of jasmine.

3.   Climate. Latitude, hence climate, influence both odor intensity and quality as in the case of lily of the valley or lilac flowers.

4.   Environment. Similarly to wine-making, soil as well as and sun exposure can quantitatively affect the fragrance of flowers. As an example, in mid-February 1992, we isolated the fragrance given off by mature mimosa inflorescences, picked simultaneously from the same tree, under different exposures; that is, in the shade or in full sunlight. Upon sniffing, it was easy to realise that the flowers smelled different. This was fully confirmed by subsequently analyzing their trapped volatiles (Table IX).

5.   Physiologic status. Changes can occur day by day during anthesis. We were able to check this by the systematic analysis of the evolution of the composition of the headspace from broom flowers (*Spartium junceum* L.), picked from the same shrub on 5 consecutive days at 8.00 a.m. (Joulain, unpublished data).

These parameters should all be taken into account when studying the fragrances of flowers. The concept of "intact vs picked" or "living vs dead" flowers is but one aspect among several others to be considered (*15*), and should not be looked upon as a decisive element.

Acknowledgments

I thank Raymond Laurent, Jean-Philippe Fourniol and Claude Noblet for their technical assistance and Mrs. Lydia Ziegler for her secretarial help.

Literature Cited

1.   Bicchi,C.; Joulain,D. *Flav. Frag. Journal*, **1990**, *5*, 131.
2.   Kaiser,R. In *Perfumes, Art, Science and Technology*; Müller, P. M.; Lamparsky, D., Eds.; Elsevier Applied Science: London and New York, **1991**, 213-250.
3.   Joulain,D. In *Progress in Essential Oil Research*; Brunke, E.-J., Ed.; Walter de Gruyter: Berlin and New York, **1966**, pp 57-67.
4.   Mookherjee, B. D.; Trenkle, R. W.; Wilson, R. A.; Zampino, M.; Sands, K. P.; Mussinan, C. J. In *Flavors and Fragrances: A World Perspective*; Lawrence, B. M.; Mookherjee, B. D.; Willis, B. J., Eds.; Elsevier Science Publishers: Amsterdam, **1988**, pp 415-424.
5.   Zhu, L.; Lu, B.; Luo, Y. *Zhiwu Xuebao*, **1984**, *26*, 189.
6.   Kaiser, R. In *Flavors and Fragrances: A World Perspective*; Lawrence, B. M.; Mookherjee, B. D.; Willis, B. J., Eds.; Elsevier Science Publishers: Amsterdam, **1988**, pp 669-684.
7.   Bu, X.; Huang, A.; Sun, Y.; Lin, M.; Wu, Z. *Acta Scientarium Naturalium*, Univ. Pekinensis, **1987**, *6*, 53.
8.   Desheng,D. *Perf. Flav.*, **1989**, *14*, Sept.-Oct., 7.
9.   Tsuneya, T.; Ikeda, N.; Shiga, M.; Ichikawa, N. In *Proceedings 7th International Congress of Essential Oils*, Kyoto (1977), **1979**, pp 454-457.
10.   Bessiere, J. M.; Pellecuer, J.; Allain, P. *Fitoterapia*, **1985**, *56*, 62.
11.   Tatsuka, K.; Suekane, S.; Sakai, Y.; Sumitani, H. *J. Agric. Food Chem.*, **1990**, *38*, 2176.
12.   Surburg, H.; Güntert, M.; Schwarze, B. *J. Ess. Oil Res.*, **1990**, *2*, 307.
13.   Brunke, E.-J.; Hammerschmidt, F.-J.; Schmaus, G. *Dragoco Report*, **1992**, *1*, 3.
14.   Altenburger, R.; Matile, P. *Planta*, **1990**, *180*, 194.
15.   Mookherjee, B. D.; Trenkle, R. W.; Wilson, R. A. *Pure Appl. Chem.*, **1990**, *62*, 1357.

RECEIVED August 19, 1992

Chapter 15

Headspace Analysis of Volatile Compounds Emitted from Various Citrus Blossoms

T. Toyoda, I. Nohara, and T. Sato

Central Research Laboratory, Takasago International Corporation, 5–36–31 Kamata, Ohta-ku, Tokyo 144, Japan

Thermal desorption is routinely used to release the adsorbed volatile compounds in the headspace analysis, but it requires so much heat to complete the desorption that it frequently causes the isomerization and/or decomposition of thermally fragile compounds. The solvent desorption method was proved to be more convenient and versatile for the component analysis and the sensory evaluation of the volatiles emitted by plants. The compositions of volatiles emitted by the blossoms of thirteen citrus species were analyzed by this method. The comparison of analytical results explained partially the close relationship of the morphological classification of these plants. On the other hand, the compositional resemblance was found between the headspace sample and the essential oil of *C. aurantium* L.

The long cherished dream of perfumers is to reconstruct the scents of natural flowers and to use them in creating fragrances. For this purpose, the oils extracted from flowers by steam distillation or solvent extraction have been used since ancient times. However, the scents of natural flowers can hardly be imagined from these oils even in the best case where the fragrances are preserved well during the extraction processes. It is very common to have extracts show completely different smells from those of the flowers and the degree of resemblance depends on the nature of the flowers themselves. Since the flowers are picked from the plants prior to distillation or extraction, the scents are overcome by the undesirable odors of exudations from the wounded parts or the plasmic components from the inner parts of petals or calyces. Headspace analysis which was very popular during the last decade was supposed to be suitable for this field, because it catches only the same volatile compounds emitted from plants that we detect by our nose, without any interference of components of the plant organs. Though the headspace analysis consists of the static and the dynamic methods, we employed the latter using the adsorption onto TENAX TA for the analysis of flower scents because of its higher sensitivity and mobility. The dynamic headspace analysis using adsorbents was

0097–6156/93/0525–0205$06.00/0

widely used in the field of air pollution analysis and many modifications were found in the literature. However, there were two problems to be solved when it was applied to the analysis of volatile compounds of plants.

Adsorption Efficiency

Glass tubes of ca 5mm i.d. are commonly used for the adsorption sampling tubes; however, they cause a serious pressure gap and the large volume of the aerial sample cannot be treated at the sampling sites in a short period with a mini sampling pump. In the field of air pollution analysis, the target of analysis can be focused onto a single compound or chemical group and it is relatively easy to avoid the influence of the background by the choice of the appropriate selective detectors and the operating conditions. In contrast, as the analysis of fragrant materials from the plants requires that almost all data be utilized, the background level must be kept as low as possible throughout the chromatographic data. It can be achieved by increasing the relative concentrations of the target sample to the background. In this case, it is essential to treat a large volume of the aerial sample effectively by apparatus with high pumping capability. The adsorption tube which is composed of a glass tube of 8mm i.d. and 400mg of TENAX gives ca. 4cm thickness of the adsorption bed. Through this tube, the mini sampling pump can afford a flow rate of ca. 1l/min which is sufficient for the purpose of our analysis. The effectiveness of the adsorption can be ascertained by sniffing the waste air at the outlet of the pump. No significant smell was perceived to pass through the adsorption system.

Improvement of the Desorption Process

At first, we used the thermal desorption method to regenerate the fragrant materials from the sampling tubes according to the general process of the air pollution analysis. The adsorption tubes were heated to 250°C which is lower by 70°C than the pretreatment temperature of TENAX, and the desorbed volatiles were transferred to gas chromatograph by helium gas flow. The volatile compounds released were then focused on the narrow part of the analytical column by the procedure described in the experimental section. In this way, we were able to achieve high sensitivity and resolution. However, the oils which were reconstructed with the chemicals on the basis of the analytical results from the thermal desorption method had completely different smells from those of the natural flowers, and were offensive chemical odors. Analytical data showed too many low boiling compounds which gave the reconstructed oil a pungent smell. This tendency may be caused by the incomplete desorption of the less-volatile compounds. The temperature must be raised to get more effective desorption, but 250°C was thought to be close to the limit of the background produced by thermal decomposition of the adsorbent. Furthermore, the thermal decomposition and/or isomerization of monoterpenoids had taken place already even at this temperature (1). Therefore, we had to abandon the method of increasing the desorption efficiency by raising the desorption temperature. Several methods appear in the literature (2) where the adsorbed samples are desorbed by the solvent extraction using carbon disulfide, dichloromethane or methanol. However, as the instrumental analysis and sniffing must be coordinated in the analysis of the fragrant materials, the solvents which possess strong smells or high boiling points should not be used for extraction. For this purpose, diethyl ether seemed to be suitable because of its high volatility and weak smell and it indeed gave the satisfactory results. The smell of the oils extracted from TENAX by ether bore a close resemblance to that of the natural flowers. It is useful not only in the component analysis using the instruments but also in the sensory evaluation in laboratories remote from the blooming site or during different seasons when the flowers are not available. Besides, there are more merits for these oils samples in which the injection volume can be controlled and one sample can be injected repeatedly unlike the thermal desorption method. By these properties it becomes possible to standardize the analytical data.

Collection of Odorous Samples from Various Citrus Flowers

After the conditions for the adsorption and the desorption were established, the volatile components of chosen related flowers were analyzed in order to investigate the relation between the morphological classification and the compositional similarity of the components emitted by them. Citrus flowers are suitable for this purpose because many closely related citrus plants are available in a small area where citrus fruits are produced. We collected the volatile samples from thirteen species of citrus flowers at Minami-Uwa, Ehime prefecture in the southwest part of Japan. The names of citrus plants used for this experiment are listed in Table I, roughly in order of their fruit size. These plants bloom almost simultaneously during the first two weeks of May in this area. All samples were collected for two hours from about noon when the flowers seem to have the most aroma, though emission of volatile compounds is perceivable all day long.

Component Analysis

The components of the volatile sample were analyzed by GC/MS. At first, the whole adsorbent of one sample was stirred to give uniformity to the sample and seven eighths of it was used for the solvent desorption and one eighth for the thermal desorption. The number of components identified amounted 174 in which the solvent desorption and the thermal desorption contribute to 155 and 165 respectively. The major components of each sample from the thermal desorption and the solvent desorption are listed in Table II and III respectively. The compounds apparently originating from wrapping materials, adsorbent, or from the environment, were removed from the lists. These included acetone, toluene, xylenes, 2-ethylhexanol, N,N-dimethylacetamide, acetophenone, diethylene glycol and phenol. The quantities of the constituents are expressed in intensities relative to the main component in the sample to clarify the relations between each constituent and to give uniformity in the data of all samples. Alloocimenes which were detected only in the thermal desorption method clearly indicate that the thermal isomerization of monoterpene hydrocarbons or thermal degradation of oxygenated monoterpenoids takes place at the desorption temperature. Furthermore, the relation of the data between these two desorption methods are

Table I. The Citrus Species Used in the Experiment

No.	Symb.[a]	Common Names	Botanical Names
1	B1	Buntan	*Citrus grandis*, Osbeck
2	B2	Kawachibankan	*C. kawachinensis* Hort.
3	M1	Natsumikan	*C. natsudaidai* Hayata
4	M2	Amanatsu	*C. natsudaidai* Hayata
5	M3	Hassaku	*C. hassaku* Hort. ex Y. Tanaka
6	M4	Kabosu	*C. sphaerocarpa* Tanaka
		or Daidai	*C. aurantium* Linn
7	M5	Iyokan	*C. iyo* Hort. ex Tanaka
8	M6	Sanbokan	*C. sulcata* Hort. ex Tanaka
9	S1	Navel	*C. sinensis* Osbeck
10	S2	Unshu	*C. unshiu* Marcovitch
11	S3	Lemon	*C. limon* Burm. f.
12	T1	Sudachi	*C. sudachi* Hort. ex Shirari
13	T2	Tachibana	*C. tachibana* Tanaka

[a] The species are listed roughly in order of their fruit size.
B, Big fruits; M, Medium; S, Small; T, Tiny.

arbitrary among the samples. It means that the thermal desorption methods could not afford the complete release of the adsorbed compounds and that the data obtained by this method have less reliability. Ten samples among thirteen contain linalool as the main component and is predominant in the volatile components detected. The important components in the volatiles of the citrus flowers other than linalool are methacrolein, myrcene, methyl anthranilate and indole which have low odor threshold and are detected from almost all samples without exception. Each of these compounds gives a characteristic odor to citrus flowers. Benzyl cyanide is detected in eleven samples and contributes to the odors, though in high concentrations it smells pungent and strongly chemical.

Compositional Similarities of Flower Volatiles

The similarity indices were calculated using the data from the solvent desorption listed in Table III by equation 1, and the results of the calculation are listed in the upper right part of Table IV.

$$\text{S.I.} = \frac{\sum a_i \cdot b_i}{\sqrt{\sum a_i^2} \cdot \sqrt{\sum b_i^2}} \quad i = 1 \sim 58 \quad (1)$$

The linalool contents are the exclusively dominant factor determining the similarity. This means that the results of the direct calculation from the raw data do not indicate the proper similarities, and they may be real when they are kept still high even after decreasing the contributory factor of linalool. The indices calculated under the condition where the contributory factor of linalool is reduced to one quarter are listed in the lower left part of Table IV. These results show the close relation of M5 with M2 and M3 and of M3 with B2. Though these data are too ephemeral and incomplete to justify a discussion on chemotaxonomy, the dendogram was drawn by the cluster analysis on the basis of similarity indices (Figure 1). In comparison with that of the morphological classification by T. Tanaka (3) (Figure 2), much agreement was found in the species belonging to *Archicitrus* (subgenus) having high compositional similarities in the upper right of Table IV while lower ones such as S2, T1 and T2 belong to *Metacitrus*. The only exception was *C. limon* Burm f.(S3) which belongs to the subgenus *Archicitrus* but the content of flower volatiles is far different from that of others of the same subgenus.

Relation with the Composition of Essential Oils

Orange flower absolute and neroli oil are commercially available essential oils from citrus flowers. Orange flower absolute is the product of the solvent extraction and the neroli obtained from steam distillation while the raw material of both oils is flowers of bitter orange (*C. aurantium* L.) whose fruits are too sour to be edible. The major components of these oils are listed in Table V which expresses the quantities again in intensities relative to the main component. The striking feature

Table II. The Major Volatile Constituents Detected in Thermal Desorption Method

	B1	B2	M1	M2	M3	M4	M5	M6	S1	S2	S3	T1	T2
Methacrolein	0.8	4				1	1		2	3			6
Pentanal					0.9					+[a]	13	3	2
Undecane					0.3			7					
Myrcene	85	19	67	57	49	35	38	58	100	100	4	100	57
3-Methyl-2-butenal	6	6	4	3	4	1	3	2	3			2	4
Limonene	8	29	8	3	8	18	3	29	8	3	81	2	38
1,8-Cineole	1			6			1				4		
2-Methyl-5-isopropenyl-2-vinyltetrahydrofuran		5											
β-Ocimenes	10	9	10	7	19	19	10	31	24	10	21	11	90
p-Cymene	9		36	7	19		9	35	12	11			
cis-3-Hexenyl acetate								3	2	1	5		
6-Methyl-5-hepten-2-one	3	4	5	4	8	4	3	9	0.5	0.5	100	1	3
Alloocimenes	2	2	1	0.8	3	2	2	3	2	0.9	2		9
Nonanal			0.4	1				0.8		0.1			
Linalool-3,6-oxides	1	6	3	0.7	3	1	2	3	3		13	3	
Decanal				0.6				0.8				5	
Pentadecane			1						2		11		10
Benzaldehyde	1	2	5	5	0.8	2	4	3	2	1	6	6	4
(2-Methyl-6-methylene-1,7-octadien-3-one)b	8	0.8	24	6	0.3			0.3	22	4	23	9	2

Compound												
Linalool	100	100	100	100	100	100	100	100	100	86	7	100
Linalyl acetate							11					
6-Methyl-3,5-heptadien-2-one		0.6	0.2		0.6	0.2	0.3	0.2	0.2	0.2		0.8
3,7-Dimethyl-1,5,7-octatrien-3-ol	2	14	8	4	4	6	4	2	2	14		6
Caryophyllene	2	14	8	4	4	6	4	2	2	14		6
Caryophyllene	2	4	11	3	4			2	2	11	23	16
3-Methylene-7-methyl-1,5-octadien-7-ol	2		10	2	1	0.4	1	0.2	10	2		2
Heptadecenes									0.4			23
3-Methylene-7-methyl-1,7-octadien-6-ol	2	1	8	1	2	0.7	0.8	0.3	9	2		2
Nerol & Geraniol	3	7	2	2	8	3	10	5	0.4	0.1		13
Phenethyl alcohol		1		2	0.3	1	7	2	2	0.3		6
Benzyl cyanide		2	2	4	0.6	1	6	4	2	1		16
3,7-Dimethyl-1,5-octadiene-3,7-diol	1	6	10	1	4	2	3	0.2	8			7
3,7-Dimethyl-1,7-octadiene-3,6-diol	0.8	9	6	0.7	4	2	2		8			7
Methyl anthranilate	0.9	3	4	6	7	2	12	1	4	2	10	0.4
Indole	0.3	1	0.4	3	0.8	5	7	2	5	0.6	10	

a Detected at the level lesser than 0.1 versus main component.
b Tentative assignment

Table III. The Major Volatile Constituents Detected in Thermal Desorption Method

	B1	B2	M1	M2	M3	M4	M5	M6	S1	S2	S3	T1	T2
Methacrolein	2	3	13	2	3	0.4	0.7	0.7	2	1			15
1-Buten-3-one	0.5	2	5	1	1	0.3	0.5	0.2	2				0.5
Sabinene	0.2		0.4	1			0.3	0.2	3	0.5			0.9
Myrcene	24		11	37	9	1	24	5	41	100		96	18
3-Methyl-2-butenal	0.5	0.4	3	0.5	0.4		0.3	0.1	2	1		13	
Limonene	5	0.8	1	1	0.7	1	0.3	0.5	2	1	5		14
β-Phellandrene	0.3			0.5					0.5	1		4	
1,8-Cineole			1	0.7			0.7						
β-Ocimenes	1	0.1		4	1	1	2	5	7	16	3	6	57
p-Cymene	1	1	7	2	2		0.6	1	1	7		17	0.3
6-Methyl-5-hepten-2-one	+[a]	0.4	0.9	0.7	0.7	0.1	0.4	0.9		0.6	9		0.5
Nonanal		+	0.1		0.1		0.1	0.1			2	13	
Myrcene-6,7-oxide	0.3		0.7		0.3				0.5	1		14	
Linalool-3,6-oxides	0.4	0.7	3	0.4	1	0.1	0.5	0.3	0.9				2
Acetic acid				0.6		0.3							2
β-Ocimene-6,7-oxides	0.3			0.2				0.1	0.5	1			0.6
Decanal		+		0.2	+	+		+			5	14	
Pentadecane			2	0.8	0.4	0.4	0.4		1	0.7	45		6
Benzaldehyde		0.3	1	0.3		0.1		0.2				14	0.1
(2-Methyl-6-methylene-1,7-octadien-3-one)[b]	5		14	4	1		1	0.5	3	12		73	
Linalool	100	100	100	100	100	100	100	100	100	+	1		100

Compound											
Linalyl acetate				11							
β-Elemene		4	0.4	1				2			
3,7-Dimethyl-1,5,7-octatrien-3-ol	0.4	6	1	0.2	0.3	0.1	0.9				0.6
Caryophyllene	0.4	10	1	0.2	0.9	0.7	19	3	11		22
3-Methylene-7-methyl-1,5-octadien-7-ol	5	11	1	+	0.9	0.2	5	11		83	0.7
β-Farnesene		0.1	0.9	0.2	0.3	0.9					
α-Terpineol	0.6	6	2	+	1	0.6	1	0.8			0.9
Heptadecane		3	1	0.2	0.4		2	0.6	13		16
Heptadecenes		4	1	0.5	3		5	0.7	9		48
Citrals	+	0.6	1		0.9	0.4			2		10
3-Methylene-7-methyl-1,7-octadien-6-ol	3	10	0.9	+	0.8	0.1	4	10		100	
α-Farnesenes	+	0.5	1	0.3	0.4	0.2			25		7
Methyl salicylate		2	0.1			+	0.7	0.1			0.5
Citronellol	+		0.3	0.1	2			0.1	2		4
2,6-Dimethyl-1,5,7-octatrien-3-ol	+				2	0.4					
Linalool-6,7-oxides	1	2	1	0.2	0.3	0.2	3	1			0.3
Caproic acid	0.8	5	3		6	2				13	
Nerol + Geraniol		1	1	+				1			12
Geranylacetone		0.3	1	+	0.7	0.4		1	8		

Continued on next page

Table III. Continued

Compound													
Phenethyl alcohol			5	4	2	2	4	1	5	2			6
Benzyl cyanide			2	2	0.7	1	2	2	2	3		8	11
Nonadecane				1		1			1				1
Jasmones				0.1	0.3	0.1	0.3	+	0.8	2			
3,7-Dimethyl-1,5-octadiene-3,7-diol	23	12	34	12	8	6	7	1	13				14
3,7-Dimethyl-7-hydroxy-1,5-octadien-3-yl-acetate						0.5							
Caryophyllene oxides	0.8	0.3	3	0.4	1	0.1			5				2
5-(3-Furyl)-2-methyl-1-penten-3-one				0.6					0.5			8	
Nerolidol			1	6	9		0.5	3	8		1		
3,7-Dimethyl-1,7-octadien-3,6-diol	15	17	34	7	9	7	7	1	12				10
Anthranilaldehyde	2	4	4	2	4	+	1	0.4	0.8		13		2
Methyl anthranilate	8	39	22	37	32	7	35	3	8	26	100		1
Phenylacetaldoxime			10			0.2		2					
Benzoic acid													
Indole	0.3	8	5	7	2	12	12	3	9	3	93		0.8
Methyl N-formylanthranilate	1		3	1	0.5	0.1	0.4	+	1	1	0.8	2	
Myristic acid				2			0.5	0.3					
Palmitic acid			5	4		0.7	0.9			0.2	36		

a Detected at the level lesser than 0.1 versus main component.
b Tentative assignment.

Table IV. Compositional Similarity Indices of Citrus Volatiles

	B1	B2	M1	M2	M3	M4	M5	M6	S1	S2	S3	T1	T2
B1	1.00	0.92	0.94	0.95	0.95	0.94	0.94	0.95	0.96	0.24	0.06	0.18	0.77
B2	0.62	1.00	0.92	0.93	0.98	0.94	0.97	0.92	0.87	0.09	0.28	0.00	0.72
M1	0.82	0.73	1.00	0.90	0.91	0.89	0.89	0.86	0.89	0.18	0.18	0.20	0.75
M2	0.80	0.76	0.68	1.00	0.96	0.90	0.99	0.90	0.96	0.40	0.27	0.22	0.75
M3	0.70	0.93	0.71	0.88	1.00	0.95	0.98	0.95	0.91	0.16	0.22	0.06	0.75
M4	0.65	0.72	0.60	0.58	0.69	1.00	0.94	0.98	0.90	0.03	0.12	0.01	0.76
M5	0.74	0.86	0.64	0.95	0.91	0.68	1.00	0.93	0.93	0.29	0.29	0.13	0.75
M6	0.67	0.59	0.48	0.62	0.70	0.82	0.67	1.00	0.91	0.07	0.05	0.03	0.78
S1	0.84	0.47	0.65	0.84	0.62	0.54	0.73	0.63	1.00	0.39	0.12	0.25	0.81
S2										1.00	0.19	0.65	0.21
S3											1.00	0.00	0.09
T1												1.00	0.09
T2	0.41	0.25	0.39	0.38	0.31	0.32	0.35	0.46	0.54				1.00

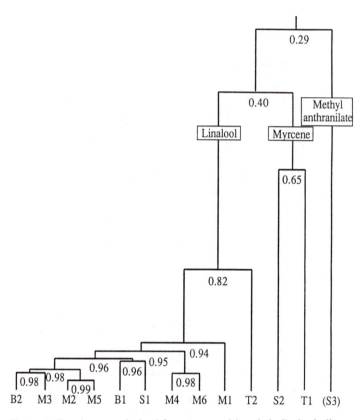

Figure 1. Dendrogram derived from compositional similarity indices.

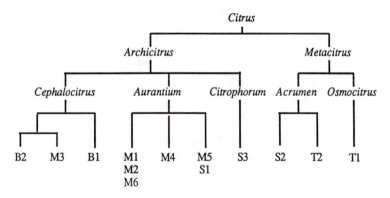

Figure 2. Dendrogram of morphological classification.

Table V. The Major Constituents of Essential Oils in Comparison with M4

	M4	Abs.	Neroli
Myrcene	1	0.3	4
Limonene	1	1	30
ß-Phellandrene			0.5
ß-Ocimenes	1	1	13
p-Cymene			0.4
Linalool-3,6-oxides	0.1	0.3	0.7
Benzaldehyde	0.1		0.2
Linalool	100	100	100
Linalyl acetate	11	15	11
Caryophyllene	0.2	0.1	1
α-Terpineol	+[a]	1	10
Nerol + Geraniol	+	0.6	9
Phenethyl alcohol	2	2	
Benzyl cyanide	1	1	0.3
Nerolidol		11	15
Methyl anthranilate	7	5	0.5
Indole	12	8	0.4

[a]Detected at the level lesser than 0.1 versus main component

of the relation between the analytical data of essential oils and the headspace samples is that M4 is the exception as it contains linalyl acetate and its ratio to linalool is almost the same as that of essential oils. Moreover, the composition of the volatile compounds of M4 is generally very similar to that of orange flower absolute. The similarity index of 0.99 was given for the composition of the major components. The distinct difference between them is that only essential oils contain α-terpineol and nerolidol as major components. Since it is frequently observed that these compounds are detected in only trace amounts in the headspace analysis while the oils extracted with solvent contain them as major components, they may be thought to come from inner parts of the flower petals or other organs of the plants. The fruits of M4, like bitter orange, are not edible due to their sourness and are used only for vinegar in areas where they are grown. The common names of M4 are used arbitrarily in local areas and farmers do not know exactly which plant is being grown. However, the comparison with orange flower absolute described above shows that *C. aurantium* L. may be the correct name for the plants which we collected for the headspace sample as M4.

Scope and Limitations

There was a good correlation between the morphological classification and the compositional analysis in this experiment. Headspace analysis of flower volatiles may help in the classification of plants.

The advantages of the solvent desorption method are that the smell of the collected samples is easily regenerated by extracting the adsorbent with a small volume of ether without any instruments, and that the odorous extracts can be applied directly to the sensory evaluation. The method is applicable wherever the sample is located because of the mobility of this apparatus. The differences in the compounds emitted by different parts of the plants can be observed. Changes of composition with time can also be followed. Multiple analyses showed much fluctuation in the concentrations of the components. It is necessary to perform many experiments and average the analytical results to arrive at general conclusions. In the production of essential oils, many plants must be processed and therefore the composition of the volatiles is an average of the individual plants.

Experimental Section

Pretreatment of Adsorbent. TENAX TA 60/80 mesh was washed thoroughly with methanol for six hours in Soxhlet extraction apparatus. After drying, it was packed into 5mm i.d. x 160mm glass tube connected to the carrier gas line and placed in the column aging oven. The volatile contaminants were cleared off by heat under the nitrogen gas flow. The temperature of the tubes was raised gradually to prevent the pulverization of granules of the adsorbent and kept at 320°C for eight hours.

Procedures for the Collection of Volatiles.
TENAX TA 60/80 mesh was transferred from the narrow tube used for pretreatment to the sampling tube of 8mm i.d. x 100mm at the sampling site. It

resulted in about a 40mm thickness of the adsorption bed. A bunch of the blossoms was enveloped in a 20*l* Tedlar bag which had been cut open at the bottom and the cut end of the bag was narrowed by binding it with adhesive tape. The sampling tube was placed between the Tedlar bag and the mini sampling pump and connected to the sampling tip of the bag and the inlet of the pump. The aerial sample was passed through for two hours at the flow rate of ca. 1*l*/min.

GC/MS Analysis. GC/MS was measured in EI mode at ionization voltage of 20eV on Hitachi M-80B mass spectrometer connected to Hewlett Packard HP-5790A gas chromatograph equipped with the chemically bonded PEG-20M capillary column of 0.25mm x 50m. The oven temperature was raised linearly from 70 to 210°C at the programmed rate of 4°C/min.

Thermal Desorption. After the TENAX TA in the sampling tube was stirred to give it uniformity, fifty milligrams of it was packed into the desorption tube of 5mm i.d. and the tube was connected to the bypass line of the carrier gas. It was necessary to dry the TENAX by passing the carrier gas at room temperature prior to the injection. The volatile compounds were then released by heating the tube to 250°C with the electric furnace and introduced into the split-less injection port by the carrier gas flow. During the desorption, about 10cm of the analytical column near the injection port were chilled by means of a dry ice acetone bath to condense the released volatiles. It required about ten minutes to transfer a sufficient amount from the adsorption tube to the analytical column. After the transfer was completed, the dry ice acetone bath was removed and the column oven was operated on the sequence of the normal analytical condition.

Solvent Desorption. Purified diethyl ether of one milliliter was poured onto the remaining TENAX TA in the sampling tube followed by the addition of 2*ml* of ether in 1*ml* portions during approximately 10 minute intervals. The ether dripping was collected and combined with that recovered by a gentle stream of nitrogen gas passed through the adsorbent. The ether solution was applied directly to the sniffing test by dipping the thin strip of filter paper into it. For GC/MS analysis, the ether solution was concentrated by gently warming at atmospheric pressure, and the residual oil was injected in the conventional way.

Literature Cited

1. Burger, B. V.; Munro, Z. *J. Chromatogr.* **1986**, *370*, 449-464.
2. Curvers, J.; Noy, Th.; Cramers, C.; Rijks, J. *J. Chromatogr.* **1984**, *289*, 171-182.
3. *Citrus Industry*; Reuther, W.; Webber, H. J.; Batchelor, L. D., Eds.; A Centennial Publication of the University of California; Division of Agricultural Sciences; CA, US, **1967**; *Vol. 1*; pp 364-367.

RECEIVED August 19, 1992

Chapter 16

Volatile Components of Apricot Flowers

Ichiro Watanabe, Osamu Takazawa, Yasuhiro Warita, and Ken-ichi Awano

Kawasaki Research Center, T. Hasegawa Company, Ltd., 335 Kariyado, Nakahara-ku, Kawasaki, Japan

Volatile components from apricot flowers were analyzed. Apricot flowers were extracted with hexane. This extract was distilled to exclude the non-volatile compounds, and the distillate was further separated by column chromatography into three fractions. These fractions were analyzed by GC and GC/MS. The compounds were identified by GC/MS, were then verified by synthesis. Eighty five compounds were identified, including lilac alcohol, lilac aldehyde, lilac acetate, 6-hydroxy-2,6-dimethyl-2,7-octadienal and (2Z)-2,6-dimethyl- 2,7-octadien-1,6-diol.

Apricot (Prunus Armeniaca L. var. Ansu Maxim.) belongs to the rose family, and is native to China. In spring, it blooms with a small light pink flower which has a pleasant fragrance. The fruit is used in jams and preservatives and is also consumed fresh. This fruit has a unique characteristic smell, about which many reports have been made with regards to its volatile components. To our knowledge, however, there is no work reported on the fragrance of its blossoms. We wish to report analytical results on the volatile components of an extract of the apricot flowers.

Experimental Procedures

<u>Materials.</u> Apricot flowers were collected in April in the Nagano Prefecture, Japan.

<u>Extraction of Apricot Flowers.</u> The amount of 3.845 kg of apricot flowers was extracted twice with 15 liters of distilled hexane. The extracts were combined and dried with anhydrous sodium sulfate. The hexane was removed by distillation at atmospheric pressure, resulting in 3.39 g yield (0.13%).

0097–6156/93/0525–0220$06.00/0
© 1993 American Chemical Society

Distillation of Extract. The extract was distilled by using a semimicro distillation apparatus with the receiver cooled with liquid nitrogen, resulting in 0.13 g yield of distillate.

Fractionation of Distillate. The distillate was separated into 3 fractions by silica gel column chromatography (Fig.1).

GC Conditions. A Shimadzu instrument, model 14A, with a FID and a 0.25 mm i.d.x 60 m fused silica capillary column coated with PEG-20M was used. The column temperature was programmed from 70°C to 220°C at 3°C/min. Injection port and detector temperature were 250°C. Nitrogen gas was used as the carrier gas at flow rate of 1ml/min.

GC-MS Conditions. A Hitachi, model M80B, mass spectrometer was used. The GC column conditions were the same as described above for the GC. Other operating parameters were as follows: column temperature, programmed from 60°C to 210°C 3°C/min, carrier gas, helium; ionizing volt, 70eV; accelerating voltage, 3000V: ion source temperature, 190°C.

GC-FTIR Conditions. The IR spectra were measured with a Hewlett Packard GC/FTIR system. The HP 5965B FTIR system was connected to a HP 5890 - series II gas chromatograph; column, fused silica capillary column (0.25 mm i.d. x 60m) coated with PEG-20M.; column temperature, 70°C for 1 min, followed by an increase to 220°C at a rate of 3°C/min: injection port temperature, 250°C; IR cell temperature, 250°C.

NMR Conditions. The NMR spectra were recorded on a JEOL FX90A spectrometer at 90MHz, tetramethyl silane (TMS) was used as internal standard with chemical shift values being expressed in (CDCl$_3$) ppm.

HPLC Conditions. A Waters instrument, model 600E, equipped with a Waters Differential Refractometer R401 detector, and a 10 mm i.d. x 25 cm x 4 Develosil 60-5 column was used. The mobile phase was hexane-ethyl acetate (70: 30, v/v). The flow-rate was 4 ml/min.

Preparation of 6-hydroxy-2,6-dimethyl-2,7-octadienal (3), lilac aldehyde (4), lilac alcohol (5) and lilac acetate (6). 6-Hydroxy- 2,6-dimethyl-,7-octadienal, lilac aldehyde and lilac alcohol were prepared by following the route previously used by Ohno (*1, 2*). Lilac acetate was prepared by the reaction of acetyl chloride with lilac aldehyde (Fig. 3).

Preparation of (2Z)-2,6-dimethyl-2,7-octadien-1,6-diol. Linalyl acetate (1) was oxidized with ozone, then reduced with acetic acid-zinc, resulting in obtaining selectively 4-acetoxy-4-methyl- 5-hexenal (7). Then (7) was treated with alpha-diethyl phosphonopropionate to yield ethyl 6-acetoxy-2,6-dimethyl-2,7-octadienoate (8). This (8) was a mixture of 2E and 2Z isomers at a ratio of 9 to 1, measured by GC. Then it (8) was treated with lithium aluminum hydride to

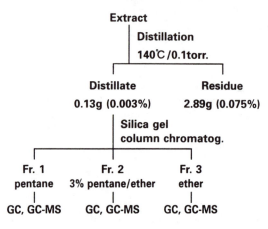

Figure 1.   Extraction and separation of *Prunus armeniaca* flower volatiles.

yield 2,6-dimethyl- 2,7-octadien-1,6-diol, which was a mixture of 2E and 2Z isomers at a ratio of 9 to 1 (Fig. 4). The 2Z isomer was isolated by HPLC.

Results and Discussion

The extract of apricot flower was distilled to obtain the volatile fractions. The distillate was separated into three fractions by silica gel column chromatography. These fractions were analyzed by GC and GC/MS. The identification of the components was made by comparing their GC retention indices and mass spectra with those of authentic compounds. The compounds, identified by GC/MS, were verified by synthesis. Fig. 2 shows a gas chromatogram of the volatile compounds found in apricot flowers.

The peaks, 76, 79, 82 and 86 represent isomers of lilac alcohol. Peak number 140 is 2,6-dimethyl-2,7-octadien-1,6-diol. The mass spectra of peaks 60, 62, 63 and 66 were similar to each other. Upon further detailed study of the fragments of these spectra, it is assumed that these substances are isomers of lilac aldehyde. Synthesized lilac aldehyde (4) was found to have 4 isomers. The mass

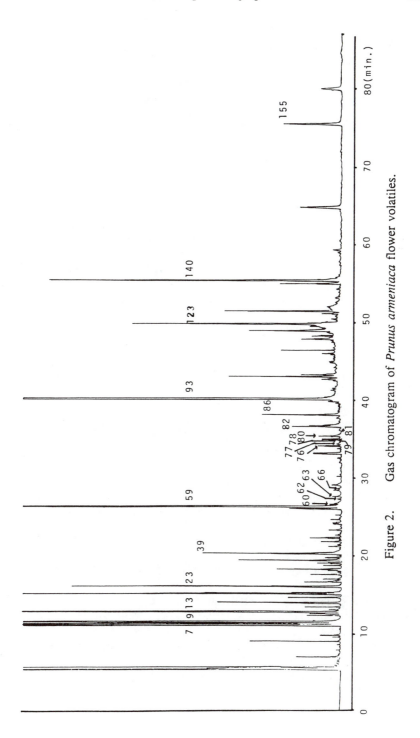

Figure 2.   Gas chromatogram of *Prunus armeniaca* flower volatiles.

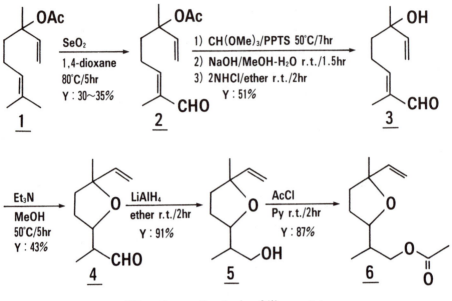

Figure 3.   Synthesis of lilac acetate.

spectra and GC retention indices were identical to that of peaks 60, 62, 63 and 66. It may be concluded that the material represented by these peaks can be identified as lilac aldehyde.

The mass spectra of peaks 77, 78, 80 and 81 were similar to each other. Mass spectrum of peak No. 77 showed: 197(4), 152(3), 137(5), 111(58), 93(40), 55(51), 43(100).   Upon detailed study of the fragments of their spectra these substances were assumed to be the isomers of lilac acetate. In the same manner as the above, synthesized lilac acetate (6) was found to have 4 isomers. Their mass spectra and GC retention indices were identical to that of peaks 77, 78, 80 and 81. Thus it was concluded that these peaks were identified as representing the isomers of lilac acetate.

The mass spectrum of peak No. 129 showed: 137(7), 119(18), 93(11), 79(18), 71(56), 55(34), 43(100).

IR spectrum showed: 3840, 3083, 2980, 2934, 1378, 1103, 1008, 925 (cm^{-1}).

Because the mass and FT-IR spectra of peak 129 were similar to that of synthetic (2E)-2,6-dimethyl-2,7-octadien-1,6-diol, it was assumed that peak 129 is be (2Z)-2,6-dimethyl-2,7-octadien-1,6-diol. The mass spectrum, IR spectrum and GC retention index of synthetic (2Z)-2,6-dimethyl-2,7-octadien-1,6-diol were identical to those of peak 129. Thus, peak 129 can be identified as (2Z)-2,6-dimethyl-2,7-octadien-1,6-diol.

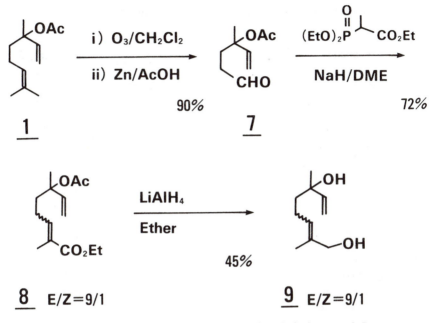

Figure 4. Synthesis of (2Z)- and (2E)-2,6-dimethyl-2,7-octadiene-1,6-diol.

The analytical results of volatile components of the apricot flower are listed in Table I. Of these components, the concentration of benzaldehyde was found to be as high as 11.0% and therefore is important in contributing to apricot flower aroma. The 6-hydroxy-2,6-dimethyl-2,7-octadienal was only a 2.2% but was found to be very important in its contribution to the apricot flower aroma. Though not a new compound, this compound was identified naturally for the first time. Lilac aldehyde has a refreshing fruity floral aroma. Lilac aldehyde and lilac alcohol have been only reported in lilac flowers (*1, 2, 3*). Lilac acetate has a fruity floral aroma. This lilac acetate is a new compound. For the unequivocal identification of the different isomers of lilac acetate, each isomer must be be synthesized and isolated in a pure state, and spectral data obtained on each pure isomer to determine the stereochemistry of each isomer.

(2Z)-2,6-dimethyl-2,7-octadien-1,6-diol is a new compound. Since the mass spectrum of peak 129b was similar to that of 6-hydroxy-2,6-dimethyl-2,7-octadien-al, peak 129b was assumed to be (2Z)-6-hydroxy-2,6-dimethyl-2,7-octadienal. It is being synthesized to verify its structure.

From our analyses, 85 compounds in the apricot flower were identified.

Table 1. Volatile Components of the Extract from Apricot Flowers

Peak Area	Compounds	GC Area (%)
2	Toluene	0.4
3	2-Methylpropanol	trace
4	Hexanal	0.1
5a	3-Methyl-3-butanol	trace
5b	3-Methyl-2-pentanol	trace
7	Ethylbenzene	15.5
8	p-Xylene	4.3
9	m-Xylene	10.3
13a	o-Xylene	6.3
13b	Heptanal	0.5
23	Octyl alcohol	1.7
29	Nonanal	1.1
31	Hexanol	0.1
32	cis-3-Hexenyl propionate	0.4
36	cis-3-Hexenol	trace
39	trans-2-Hexenol	0.1
46	1-Octen-3-ol	trace
52	4,6,6-Trimethylheptan-2-ol	trace
54	Pentadecane	trace
58	Camphor	0.4
59a	Benzaldehyde	11.0
59b	Linalool	0.1
60	Lilac aldehyde (isomer 1)	0.1
62	Lilac aldehyde (isomer 2)	trace
63	Lilac aldehyde (isomer 3)	trace
66a	Lilac aldehyde (isomer 4)	trace
66b	Hexadecane	trace
67	Undecenal	0.1
71	trans-2-Decenal	trace
72	Benzyl formate	trace
73	Heptadecane	trace
75	γ-Hexalactone	trace
76a	Benzyl acetate	0.1
76b	Lilac alcohol (isomer 1)	0.1
77	Lilac acetate (isomer 1)	trace
78	Lilac acetate (isomer 2)	trace
79	Lilac alcohol (isomer 2)	0.2
80	Lilac acetate (isomer 3)	trace
81	Lilac acetate (isomer 4)	trace
82a	Methyl salicylate	0.4
82b	Lilac alcohol (isomer 3)	0.4
83	Octadecane	0.4

Table 1. Continued

Peak Area	Compounds	GC Area (%)
85	Tridecanal	trace
86	Lilac alcohol (isomer 4)	0.7
91	Guaiacol	trace
92	Benzyl butyrate	0.1
93	Benzyl alcohol	16.7
94	Nonadecane	trace
97	2-Phenylethyl alcohol	trace
98	γ-Octalactone	trace
100a	Tetradecanal	0.2
100b	Benzyl cyanide	trace
101	Heptanoic acid	0.1
102	4-Methylguaiacol	0.3
105	Eicosane	trace
107	β-ionone epoxide	trace
110	Nerolidol	trace
111	4-ethylguaiacol	0.1
112	3-Phenylpropyl alcohol	trace
113a	2-Ethylphenol	trace
113b	Benzyl hexanoate	trace
114	2,5-Dimethylphenol	trace
115	p-Cresol	trace
116a	m-Cresol	trace
116b	Hexyl benzoate	trace
117a	Octanoic acid	trace
117b	Heneicosane	0.4
120	4-Propylguaiacol	trace
121a	Benzyl tiglate	0.9
121b	6,10,14-Trimethylpentadecan-2-one	0.3
123	6-Hydroxy-2,6-dimethyl-2,7-octadienal	2.2
125	4-Isopropylphenol	0.3
127	Eugenol	trace
128	Docosane	0.2
129a	Elemicin	0.9
129b	(2Z)-2,6-Dimethyl-2,7-octadien-1,6-diol	trace
130	Nonanoic acid	0.3
137	Jasmin lactone	trace
138a	Tricosane	0.7
138b	Cinnamyl alcohol	0.1
140	(2E)-2,6-dimethyl-2,7-octadien-1,6-diol	2.9
144	2,6-Dimethoxy-4-methylphenol	trace
145	Octadecanal	trace
148	Tetracosane	trace
155	Benzyl benzoate	1.1

Literature Cited

1. Wakayama, S.; Namba, S.; Ohno, M. *Nippon Kagaku Zasshi*, **1971**, 256-259.
2. Wakayama, S.; Namba, S.; Ohno, M. *Bull. Chem. Soc. Japan*, **1970**, *43*, 3319.
3. Wakayama, S.; Namba, S.; Ohno, M. *Bull. Chem. Soc. Japan*, **1973**, *46*, 3181-3187.

RECEIVED October 9, 1992

Chapter 17

Volatile Compounds from Strawberry Foliage and Flowers

Air versus Nitrogen Entrainment; Quantitative Changes and Plant Development

T. R. Hamilton-Kemp[1], J. H. Loughrin[1], R. A. Andersen[2], and J. G. Rodriguez[3]

[1]Department of Horticulture, University of Kentucky, Lexington, KY 40546
[2]Agricultural Research Service, U.S. Department of Agriculture, Department of Agronomy, University of Kentucky, Lexington, KY 40546
[3]Department of Entomology, University of Kentucky, Lexington, KY 40546

Headspace compounds were isolated from detached strawberry foliage by using both air and nitrogen as entrainment gases and trapping on the porous polymer Tenax. Compounds were eluted from traps with hexane, analyzed by GC and GC-MS, and identified by comparison with authentic standards. The profile of volatiles entrained with nitrogen differed considerably from that obtained with air; the former yielded more aliphatic alcohols, esters, and aromatics and the latter yielded greater quantities of terpene hydrocarbons. Air was selected for subsequent experiments including entrainment of flower headspace volatiles which contained aromatic compounds not found in foliage. The developmental stage of the plants also affected the types and quantities of compounds obtained. Considerably more headspace volatiles were isolated from foliage of flowering plants in early Spring than were obtained during later stages of plant development. Removal of the fruit which represented a metabolic sink for the plants did not significantly alter emission of foliage volatiles compared to that from controls.

Strawberry foliage is susceptible to the mite *Tetranychus urticae* Koch in early Spring at the time of flowering but develops resistance which increases through the fruiting period and post-harvest (*1,2*). As part of a study of the possible role of volatile compounds in the interaction of plants with *T. urticae* we have analyzed the headspace vapors emitted by foliage (*2,3*). This review covers the chemical aspects of the studies relating to the isolation, identification, and quantitation of headspace compounds in strawberry plants. The entomological component of this research represents a separate study and is not included here.

0097–6156/93/0525–0229$06.00/0
© 1993 American Chemical Society

We investigated the types of volatiles emitted by detached leaves placed in a flask and swept with a stream of entrainment gas which subsequently passed through a Tenax trap. Comparisons were made among the volatiles obtained using two entrainment gases, nitrogen and air. Subsequently, air was selected for further experiments and comparisons were made among the volatiles emitted at different periods during the growing season including flowering, fruiting, and after fruit harvest. Comparisons were also made between plants which were allowed to progress through the fruiting cycle normally and those from which flowers and fruit were removed to eliminate metabolic sinks which might alter the chemistry and resistance of the plants. Finally, volatiles emitted from strawberry flowers (4) were studied for chemical comparisons with those emitted by the foliage of this species.

Experimental

Strawberry (*Fragaria x annanasa* Duch.) plants cultivar 'Redchief' were grown in field plots at the University of Kentucky farm in Lexington. Trifoliate leaves or flowers were collected and transported to the laboratory for analysis within 2 hours of harvest.

Headspace Apparatus. Headspace vapors were obtained by placing approximately a 100 g sample of leaves or flowers into a 5 L round bottom flask immersed in a water bath at 30°C. The flask was fitted with a side arm adaptor connected to a 1 cm diameter glass tube packed with 1.5 g of 60-80 mesh Tenax. The Tenax trap had been rinsed with hexane and conditioned at 250°C using nitrogen at a flow rate of 500 ml/min for 1 hr.

    Entrainment gas (either air or nitrogen) from a commercial cylinder was passed through a molecular sieve filter, into the flask containing the plant tissue, and then through the Tenax trap at a flow rate of 500 ml/min. After a 20 hour trapping period the Tenax was removed from the system and the adsorbed volatile compounds were eluted with 30 ml of hexane. The hexane solution was concentrated to approximately 1 ml using a microstill equipped with a Vigreux column.

GC and GC-MS of Headspace Compounds. Approximately 3 µl of a solution of plant volatiles was injected into a Hewlett-Packard 5880A GC containing a 60 m x 0.32 mm Supelcowax 10 [poly(ethylene glycol)] column. The following chromatographic conditions were used: inlet, 220°C; column 60°C for 1 min and then 3°C/min to 220°C; FID, 240°C; He carrier linear flow rate, 30 cm/sec.

    Compounds were analyzed using a Hewlett Packard 5985A GC-MS instrument fitted with the same column as described above and operated at 70 eV in the EI mode. Mass spectra of plant components were compared to those obtained from authentic compounds and identifications were confirmed by co-chromatography of plant components with authentic compounds unless noted otherwise.

    Authentic samples for comparison with strawberry plant volatiles were gifts from Bedoukian, Inc. Danbury, CT or were purchased from commercial sources. β-Ocimene isomers were obtained from clover foliage (5); α-farnesene and germacrene D were isolated from ylang ylang oil (6).

Quantitation. Compounds were quantitated by adding cumene as an internal standard to the hexane concentrate described above and injecting an aliquot onto the GC. Peak areas were integrated electronically and mean values from three separate analyses (corresponding to three headspace sampling experiments) were calculated. Yields of compounds were determined based on fresh weight of foliage or flowers.

Results and Discussion

Foliage Components. Chromatograms of the headspace samples from strawberry foliage using both air and nitrogen as entrainment gases are shown in Figure 1. The compounds identified in the samples are presented in Table I and can be classified into three categories, aliphatics, terpenoids, and aromatics. Many of the strawberry components have been identified as headspace volatiles from other species. Among the more unusual compounds in the present study were the esters, (Z)-3-hexenyl 2-methylbutanoate and (Z)-3-hexenyl tiglate, derived from leaf alcohol, (Z)-3-hexen-1-ol. The appearance of these compounds probably indicates that strawberry foliage contained relatively large amounts of the precursors, 2-methylbutanoic and tiglic [(E)-2-methyl-2-butenoic] acid. The sesquiterpene, germacrene D, has not been widely reported in plant headspace volatiles although Buttery *et al.* (*6*) identified this compound in fig and walnut leaf volatiles. Headspace samples of plant tissues frequently contain sesquiterpene hydrocarbons and there is such a wide array of possible structures of these compounds, it is not surprising that different species emit different forms of these terpenes.

Air-versus Nitrogen Entrainment. A comparison of the average quantities of the compounds obtained from foliage collected in mid-June, using both nitrogen and air entrainment, is also presented in Table I. The comparison of nitrogen and air was undertaken since the possibility of oxidation of volatile compounds under an air stream was considered a potential chemical problem in headspace analysis of foliage.

The results presented in Table I demonstrate that both gases yielded aliphatic compounds including those derived from the lipoxygenase pathway (*7*) such as (Z)-3-hexen-1-ol, known as leaf alcohol. However, nitrogen entrainment yielded over a 100-fold greater amount of (Z)-3-hexen-1-ol and also greater quantities of several ester derivatives of this compound. Two additional alcohols, 6-methyl-5-hepten-1-ol and 1-octen-3-ol, not derived from the lipoxygenase pathway, as well as the terpene alcohols, linalool and α-terpineol, were also entrained with nitrogen. The presence of relatively high amounts of alcohols and some corresponding esters among the nitrogen entrained compounds suggests that the anaerobic conditions favored increased alcohol dehydrogenase activity in strawberry foliage. Enhanced alcohol dehydrogenase activity has been observed in plant tissues under anaerobic conditions (*8*). Nitrogen also favored the production of relatively large amounts of aromatic alcohols and esters which were not detected by air entrainment of foliage (but were isolated from flowers as noted below). These compounds are thought to be formed through the phenylpropanoid pathway via β-oxidation of cinnamoyl CoA (*9*). In contrast, air entrainment yielded monoterpene and sesquiterpene

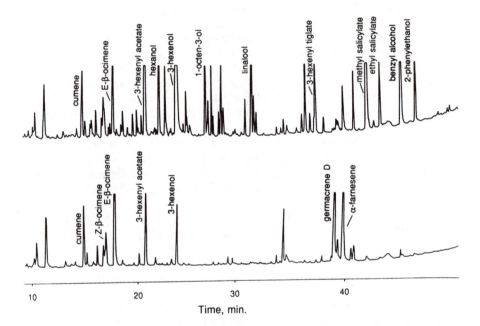

Figure 1.   Chromatograms of strawberry foliage headspace volatiles entrained with nitrogen (top) and air (bottom).

Table I.   Compounds Identified in Strawberry Foliage Headspace and Yields from Nitrogen- and Air-Entrainment

Compound	MS peaks[2]	Yield from foliage, ppb[1] Air	Nitrogen
Aliphatics			
(E)-2-hexenal	41,55,69,42,57,83		det[3]
1-hexanol	56,43,55,41,42,55	det	120
(E)-2-hexen-1-ol	57,41,43,82,67,44		det
(Z)-3-hexen-1-ol	41,67,55,82,42,69	24a	2600b
hexyl acetate	43,56,61,84,55,69	det	8
(E)-2-hexenyl acetate	43,67,82,41,100,57	det	
(Z)-3-hexenyl acetate	43,67,82,41,54,55	70a	320b
(Z)-3-hexenyl 2-methylbutanoate	67,82,57,85,41,55	det	40
(Z)-3-hexenyl tiglate	67,82,83,55,41,53	det	30
6-methyl-5-hepten-1-ol	41,95,45,69,43,67		43
1-octanol	41,56,55,43,42,70		det
1-octen-3-ol	57,43,41,72,55,85		100
Terpenoids			
(E)-β-ocimene	93,91,41,79,77,80	380a	42b
(Z)-β-ocimene	93,92,91,79,41,77	15a	4a
linalool	71,41,43,93,55,69		130
α-terpineol	59,93,43,121,81,136		8
germacrene D	161,105,91,41,79,81	250	
α-farnesene	41,93,69,55,79,107	160	
Aromatics			
benzyl alcohol	79,108,77,107,91,78	2a	300b
ethyl salicylate	120,92,166,121,65,64		40
methyl salicylate	120,92,152,121,93,65		340
2-phenylethanol	91,92,122,65,77,78		53

[1]Means (three determinations) in each row followed by different letters are significantly   different (P>0.05) according to Duncan's multiple-range test.
[2]Ions above m/z 40 in decreasing order of intensity.
[3]Compounds were not present in measureable amounts during mid-June when direct comparisons were made between yields obtained with air versus nitrogen, however, they were detected in foliage collected on other sampling dates.

hydrocarbons which were either not detected or present in significantly lower quantities in nitrogen. This marked difference between air and nitrogen in the recovery of terpene hydrocarbons was unexpected and appears more difficult to explain from a biochemical perspective. Perhaps these compounds were metabolized rapidly or were not generated in an anaerobic environment.

Buttery *et al.* (*10*) recently showed that a wide variety of synthetic volatiles including air oxidizable compounds such as aldehydes were recovered in relatively good yields using an air entrainment, Tenax trapping method. However, some degree of oxidative loss of volatiles from plant samples may occur during air entrainment. In our studies, the anaerobic conditions created when nitrogen was used for entrainment seem likely to have altered the metabolism of the foliage during the sampling process. This conclusion was based on the observation that alcohols and their corresponding esters were the main components which increased during nitrogen entrainment and that alcohol dehydrogenase, an enzyme associated with alcohol production, has been reported to increase in plants during anaerobic stress (*8*). The marked difference in composition of the headspace observed for strawberry foliage with different entrainment gases does not necessarily mean that similar results will be observed with other species. In contrast to the results with strawberry, Buttery *et al.* (*11*) used both air and nitrogen to entrain volatiles from tomato leaves and found no noticeable differences among compounds obtained with the two gases. However, based on the differences obtained with strawberry leaves using air and nitrogen and considering the anaerobic conditions created during nitrogen entrainment it was concluded that air was the preferable entrainment gas for this species and it was used in subsequent experiments.

It is noteworthy that in previous studies using steam distillation to isolate volatiles from strawberry foliage (*12*) we isolated the phenylpropanoid derivative methyl salicylate and aliphatic alcohols as were obtained with the nitrogen entrainment method. Prior to the distillation procedure the still was flushed with nitrogen to provide an inert environment and in this way the atmosphere created was similar to that in the nitrogen entrainment procedure.

Flower Components. Following analysis of the vegetative tissue headspace samples, the compounds emitted from strawberry flowers were examined for comparison (Table II). Several of the compounds produced by the flowers, especially lipoxygenase pathway products were also found in the foliage (Table I). A major sesquiterpene hydrocarbon isolated from flowers, but not detected in foliage, was tentatively identified as α-muurolene based on mass spectral data. The flowers contained several compounds produced by the phenylpropanoid pathway which, as noted above (Table I), were not detected in emissions from foliage entrained with air. Among the identified floral aromatic compounds was anisaldehyde which had an aroma reminiscent of strawberry flowers.

It is interesting that the yield of total identified volatiles from flowers was only 0.1 ppm whereas that from foliage at flowering was ca. 3 ppm. In several studies, flowers of various species yielded much greater quantities of volatiles than were emitted from strawberry flowers (*13-15*). Typically the quantity of volatiles obtained from flowers exceeds that emitted from leaf tissue (*5,16*).

Table II. Identification and Quantitation of Air-Entrained
Headspace Compounds from Strawberry Flowers

		Yield from	Detected in
1-hexanol		4	X
(Z)-3-hexen-1-ol		4	X
hexyl acetate		tr[3]	X
(Z)-3-hexenyl acetate		25	X
limonene[4]	93,68,67,136,79,94		
(E)-β-ocimene		2	X
(Z)-β-ocimene		tr	X
germacrene D		6	X
α-muurolene[5]	105,161,204,93,94,91	14	
p-anisaldehyde	135,136,77,107,92,64	tr	
benzaldehyde	106,105,77,51,78,50	7	
benzyl alcohol		3	X
methyl salicylate		2	
2-phenylethanol		tr	

[1]Ions above m/z 40 in decreasing order of intensity; data for remaining compounds presented in Table I.
[2]Means of three determinations.
[3]Integrated less than 1 ppb.
[4]Contaminant interfered with integration
[5]Tentatively identified with MS only.

The aromatic hydrocarbon 1,2-dihydro-1,1,6-trimethylnaphthalene was isolated as a major component of foliage distillates previously (17); however, this compound was not detected in foliage or flowers using the headspace trapping method. This hydrocarbon has been reported in distillates of wine and recent studies by Winterhalter (18) show that it can be formed by heating a C_{13} isoprenoid present as a glycoside in wine. Thus the elevated temperature used in the distillation process may account for the production or release of the hydrocarbon in strawberry noted previously.

Seasonal Changes and Metabolic Sinks. The quantities of volatiles obtained from foliage were determined at three periods during development including during flowering in late April, fruit development in mid-May, and approximately two weeks after fruit harvest in mid-June. Concurrently, a group of plants were not permitted to develop fruit, that is, flowers were removed from the plants as they formed. The headspace compounds from these plants were also studied in mid-May and mid-June. The purpose of this latter experiment was to determine if the removal of a metabolic sink (that is, the fruit) from strawberry plants altered the metabolism of the foliage and hence the emission of volatiles and a possible associated resistance to *T. urticae*.

The results obtained for the emission of foliage volatiles at the three seasonal periods are presented in Table III. It is clear that the foliage from plants at flowering emitted significantly greater amounts of volatiles than were emitted from foliage sampled at later periods. The amounts of lipoxygenase pathway compounds decreased sharply after flowering, but terpenoids did not show any consistent pattern of change. Comparison of the control (normal fruiting) and deflowered (non-fruiting) plants did not show marked differences in headspace volatiles from foliage. Hence the presence of the fruit as a metabolic sink did not appear to alter the foliage chemistry sufficiently to significantly change the profile of emission of volatile compounds.

Mookherjee *et al.* (19) have performed a series of experiments in which they compared emissions from intact and detached tissues and reported differences. During investigations of circadian rhythms of volatile emissions from *Nicotiana* species (15,20) we found similar patterns of emission from intact and detached flowers, however, there were difficulties associated with enclosing sufficient tissue (especially foliage) in a sampling apparatus to obtain the quantity of compounds required to perform thorough compositional studies. Future advances may permit the thorough analysis of components emitted *in situ* from plants which produce relatively small amounts of volatile compounds.

Summary

The present experiments on the emission of volatile compounds from strawberry foliage showed that very different profiles of volatiles were obtained depending on whether nitrogen or air was used as the entrainment gas. Aliphatic alcohols, their derivatives, and phenylpropanoid derivatives predominated in emissions entrained with nitrogen, whereas, terpene hydrocarbons predominated with the air entrainment

Table III. Comparison of Yields of Air-Entrained Headspace Volatiles from Strawberry Foliage at Various Periods in Control (Fruiting) and Deflowered (Non-Fruiting) Plants

Compound	Yield from foliage, ppb[1]				
	Late April flowering	Mid-May control	Mid-May deflowered	Mid-June control	Mid-June deflowered
1-hexanol	5a	tr[2]	2b	tr	tr
(Z)-3-hexen-1-ol	110a	35b	56ab	40b	59ab
hexyl acetate	21a	1b	3b	tr	tr
(E)-2-hexenyl acetate	19a		1b	tr	
(Z)-3-hexenyl acetate	2500a	470bc	620b	160c	220c
(Z)-3-hexenyl 2-methylbutanoate	9				
(Z)-3-hexenyl tiglate	54a		2b		
(E)-β-ocimene	68a	7b	44ab	120a	75a
(Z)-β-ocimene	15a	tr	7b	7b	2b
germacrene D	35ab	1b	17ab	62a	21ab
α-farnesene	190a	5b	74b	95b	40b

[1]Means in each row followed by the same letter are not significantly different(P>0.05) according to Duncan's multiple- range test. Flowering plants sampled twice; remaining plants sampled three times.
[2]Integrated yield less than 1 ppb.
(Reproduced with permission from Ref. 2. Copyright 1989, Plenum Publishing Corp.)

method. Headspace volatiles from strawberry flowers entrained with air contained phenylpropanoid derivatives which were not obtained with air entrainment of volatiles from foliage. Comparisons of the quantities of foliage compounds from plants sampled at flowering, during fruit development, and post-harvest showed that the amounts of compounds obtained were relatively high at flowering but declined significantly in the plants after this period. Removal of the fruit (metabolic sink) from the plants did not significantly change the profile of emission of volatile compounds from the foliage.

Acknowledgments

The authors thank D.D. Archbold for plant material, W.N. Einolf and S. Hoffmann for α-muurolene spectra, and Pam Wingate for manuscript preparation.

Literature Cited

1.   Dabrowski, Z.T.; Rodriguez, J.G.; Chaplin, C.E. *J. Econ. Entomol.* **1971**, *64*, 806-809.

2.   Hamilton-Kemp, T.R.; Rodriguez, J.G.; Archbold, D.D.; Andersen, R.A.; Loughrin, J.H.; Patterson, C.G.; Lowry, S.R. *J. Chem. Ecol.* **1989**, *15*, 1465-1473.

3.   Hamilton-Kemp, T.R.; Andersen, R.A.; Rodriguez, J.G.; Loughrin, J.H.; Paterson, C.G. *J. Chem. Ecol.* **1988**, *14*, 789-796.

4.   Hamilton-Kemp, T.R.; Loughrin, J.H.; Andersen, R.A. *Phytochemistry.* **1990**, *29*, 2847-2848.

5.   Buttery, R.G.; Kamm, J.A.; Ling, L.C. *J. Agric. Food Chem.* **1984**, *32*, 254-256.

6.   Buttery, R.G.; Flath, R.A.; Mon, T.R.; Ling, L.C. *J. Agric. Food Chem.* **1986**, 34, 820-822.

7.   Hatanaka, A.; Kajiwara, T.; Sekiya, J. *Chem. Phys. Lipids* **1987**, *44*, 341-361.

8.   Davies, D.D. In *The Biochemistry of Plants: A Comprehensive Treatise*; Stumpf, P.K.; Conn., E.E. Eds.; Academic Press: New York, **1980**; Vol.2; pp.581-611.

9.   Luckner, M. *Secondary Metabolism in Plants and Animals*; Academic Press: New York, **1973**.

10.   Buttery, R.G.; Teranishi, R; Ling, L.C.; Turnbaugh, J.G. *J. Agric. Food Chem.* **1990**, *38*, 336-340.

11.   Buttery, R.G.; Ling, L.C.; Light, D.M. *J. Agric. Food Chem.* **1987**, *35*, 1039-1042.

12.   Kemp, T.R.; Stoltz, L.P.; Smith, W.T., Jr.; Chaplin, C.E. *Proc. Amer. Soc. Hort. Sci.* **1968**, *93*, 334-339.

13.   Matile, P.; Altenburger, R. *Planta* **1988**, *174*, 242-247.

14.   Patt, J.M.; Rhoades, D.F.; Corkill, J.A. *Phytochemistry* **1988**, *27*, 91-95.

15.   Loughrin, J.H.; Hamilton-Kemp, T.R.; Andersen, R.A.; Hildebrand, D.F. *Phytochemistry* **1990**, *29*, 2473-2477.

16. Loughrin, J.H.; Hamilton-Kemp, T.R.; Andersen, R.A.; Hildebrand, D.F. *J. Agric. Food Chem.* **1990**, *38*, 455-460.
17. Stoltz, L.P.; Kemp, T.R.; Smith, W.O., Jr.; Smith, W.T., Jr.; Chaplin, C.E. *Phytochemistry* **1970**, *9*, 1157-1158.
18. Winterhalter, P. *J. Agric. Food Chem.* **1991**, *39*, 1825-1829.
19. Mookherjee, B.D.; Wilson, R.A.; Trenkle, R.W.; Zampino, M.J.; Sands, K.P. In *Flavor Chemistry*; Teranishi, R.; Buttery, R.G.; Shahidi, F. Eds.; ACS Symposium Series No.388; American Chemical Society: Washington, DC, **1986**; pp. 176-187.
20. Loughrin, J.H.; Hamilton-Kemp, T.R.; Andersen, R.A.; Hildebrand, D.F. *Physiol. Plant.* **1991**, *83*, 492-496.

RECEIVED August 19, 1992

Chapter 18

On the Scent of Orchids

Roman A. J. Kaiser

Givaudan-Roure Research, Ltd., Ueberlandstrasse 138, CH–8600 Duebendorf, Switzerland

The orchid family is most likely the largest and at the same time the youngest family of flowering plants. Over 25,000 species, subdivided into some 750 genera, are known to exist, and no other plant family seems to offer such a wide variety of shapes, colors and, above all, scents.

The present paper tries to illustrate with a series of selected examples that the enormous diversity in the scent composition of orchid flowers may be considered as a reflection of the equally extreme diversity in pollination principles found within this family. Discussed are also experimental aspects to be considered for the trapping and investigation of such flower scents. Finally, special emphasis is given to the identification and synthesis of new or unusual natural products, which are not only of olfactory but possibly also of semiochemical interest.

If a person is interested in flower scents, he/she will choose after a few years "The scent of orchids" as a main topic, because every imaginable scent may be found within this family. This statement is based on the olfactory evaluation of around 2200 naturally occurring species performed by the author in the course of the past ten years, from which the scents of around 250 representatives have also been trapped and subsequently investigated by analytical means. All the more surprising then, that one should often encounter the mistaken belief that orchids are scentless. This erroneous view is doubtless due to all the man-made hybrids, available today from florists, which are all too often scentless or else only weakly scented. However, of these 2200 natural species olfactorily evaluated, at least 50 % may be classified as moderately to strongly scented whereas only 15 - 20 % proved to be scentless. Whether these particular orchids appear also "scentless" to their pollinating animals is quite another matter. As will be discussed in more detail, however, many of the scented orchids show a pronounced time-dependence in their fragrance emanation. As a consequence, they have to be valuated and investigated, respectively, at the right time of day or night.

0097–6156/93/0525–0240$08.25/0
© 1993 American Chemical Society

Introduction

The scents of flowers are more than of aesthetic interest and importance; they are directly involved in the preservation of the respective plant species. This cognition was already expressed in Charles Darwin's famous book "The Various Contrivances by which Orchids are Fertilized by Insects" (*1*) and in the corresponding standard work of modern time - "Orchid Flowers : Their Pollination and Evolution" by L. van der Pijl & C.H. Dodson (*2*) - the flower scent is fully considered as part of the respective pollination principle. The analytical investigations described in literature so far have been performed in this context and, therefore, it is not surprising that olfactory aspects have hardly been discussed. Referring to the number of papers, the investigation of the scent emitted by neotropic orchids characterized by the so-called "male euglossine syndrome" attracted special interest. Male euglossine bees visit these orchid flowers not to be rewarded with food, but to collect the extremely intense floral fragrances and to use them in their own reproductive biology, probably as precursors of their own sex pheromones. It was just this highly interesting and complex plant - insect relationship studied by Dodson, Vogel, Dressler, Williams and others, that gave the stimulus for very early applications of headspace trapping techniques to the investigation of flower scents. A comprehensive literature survey concerning orchid scent and male euglossine bees up to 1982 was given by Williams (*3,4*) and Williams & Whitten (*5*). In more recent papers dealing with the same main topic Whitten, Williams & co-authors described *trans*-carvone epoxide as scent component of several *Catasetum* species (*6,7*) and ipsdienol as that of *Coryanthes* species (*8*), whereas Gerlach & Schill (*9*) used the fragrance analyses as an aid to discuss taxonomic relationships of the genus *Coryanthes*. Finally, the same authors (*10*) gave a survey on the scent composition of many additional euglossine orchids.

Starting around the same time, the equally fascinating pollination biology of various European species of the genera *Ophrys, Cypripedium, Platanthera* and others have been studied by Kullenberg, Bergström, Nilsson and other European scientists. Some very impressive examples of the role of floral fragrances as cues in animal behavior come from their work. Additionally, they have developed various techniques for studying the minute amounts of odor produced by these orchid flowers and by the insects acting as pollinators. The results of these investigations as well as the trapping techniques applied are summerized in a review by Williams (*4*) and in the more recent reviews by Bicchi & Joulain (*11*), Kaiser (*12*) and Dobson (*13*). The main topic of these European research groups was the investigation of the fascinating relationship between *Ophrys* flower and pollinator, which ensures the preservation of the around 30 *Ophrys* species, occurring mainly in the Mediterranean regions. These flowers, which are only visited by males of several genera of Hymenoptera, produce no nectar and the pollen is not available as food. According to the findings of Kullenberg and his research group, however, the copulation instinct of the insect is aroused by the flower scent, the approach flight to the flower is provoked by scent-carrying air currents, and after landing the copulation behavior is stimulated by the superficial structures and the shape of the labellum. The newest results show a certain congruency between *Ophrys* scent and the secretion of the Dufour glands isolated from the female of the respective pollinator, which confirms at least partially Kullenberg's hypothesis of a chemical mimicry of the female sex pheromone. A very recent review by Borg - Karlson (*14*) gives a comprehensive insight into this theme. Analytical investigations of orchid scents not cited here may be found in the reviews mentioned above (*3 - 5, 11 - 13*).

Trapping Technique Applied in our Investigations

In the following, the trapping technique applied in our investigation of orchid species - sorption on charcoal followed by solvent extraction - is briefly discussed.

Although activated charcoal is an extremely powerful adsorbent, which possesses optimal thermal and chemical stability and which is not deactivated by water, it was only used in a few cases up to the mid-seventies to trap the volatiles of flavor substrates. Incomplete recoveries by solvent extraction or thermal decomposition by applying heat desorption have been the main fears which, however, have been greatly overcome by the development of small traps by Grob (*15*) and Grob & Zürcher (*16*) holding only 1 - 2 mg of charcoal. The adsorbed volatiles can easily be recovered by extraction with 10 - 50 μl of carbon disulfide thus avoiding a further concentration step prior to the investigation by GC/MS. The successful preparation, handling and extraction of such traps is thoroughly described by the same authors (*15, 16*).

To collect the flower scent of a particular orchid species, the single flower or the whole inflorescence is placed in a glass vessel of adapted size and shape without damaging the flower or plant (Figure 1a.). This precautionary measure prevents additional dilution of the scent by the natural air circulation. The scented air surrounding the flower is then drawn through the adsorption trap by means of a battery operated pump over a period of 30 min. to 2 hours (30 - 60 ml/min.). The traps employed are directly attached to or placed within the glass vessel and contain a thin layer of 5 mg of charcoal embedded between two grids fused into the wall of a glass tube of 65 mm x 3 mm (available from Brechbühler AG, 8952 Schlieren, Switzerland). In certain cases, miniaturized filters containing only 1 - 2 mg of charcoal have been employed. Depending on the individual species, amounts ranging from 1 to 200 μg are collected during the usual trapping period. A second adsorption trap, connected in series, assures whether or not all the scent components were actually adsorbed in the first trap.

In the case of flowers with very complex shapes, as *Angraecum sesquipedale* (Figure 1b.), it is more practical simply to isolate the flower from the environment by, for example, a suitably shaped glass funnel. The adsorption trap is then centred as near as possible to the part of the flower where the scent release is at its maximum.

Back in the laboratory, the adsorbed scent can be recovered from the trap by extraction with 10 - 50 μl of carbon disulfide and the sample thus obtained investigated by GC/MS including complementary methods. For complete recovery of scents containing extremely polar compounds, further extraction with 10 - 50 μl of ethanol may be required.

Much attention has to be given not only to optimal experimental conditions during the investigation of the trapped scent, but also to trace constituents hardly or not characterizable by GC/MS which, however, might be of significant importance to the total fragrance. This means that every investigation of a trapped scent should be completed by a careful olfactory evaluation of all peaks eluted from the capillary columns of appropriate polarities, finally resulting in a comprehensive olfactogram.

Biological Aspects to be Considered

Being aware that the quality and quantity of the scent emitted by a defined species of a flower/plant may strongly depend on phenomena such as maturity, biological rhythm and environmental factors, the experimental parameters of headspace trapping have to be very flexibly adapted to the individual cases. Investigating orchid scents, the consideration of diurnal or nocturnal changes in quality and quantity and, correspondingly the choice of an appropriate trapping period may be very important. As an illustration, the flowers of *Epidendrum ciliare* native to Central America and the West Indies are practically scentless during day. After sunset, however, they start to emit their characteristic "white-floral" fragrance, which contains to over 50% linalool and as major constituents (E)-ocimene 6(7)-epoxide, (E,E)-4,8,12-trimethyl-1,3,7,11-tridecatetraene (12), 2-methylbutyraldoxime (2, E:Z ~ 3:1) and (E)-ocimene. As can be seen in Figure 2, the absolute amount of volatiles increases very strongly during the evening hours to the maximum between 10 p.m. and midnight and decreases afterwards to approach the zero-level in the later morning of the subsequent day. During the following night, the same pattern is recognizable, although with reduced intensity.

It also has to be taken into consideration that a defined species of a flowering plant may still show a considerable variation in the scent composition from plant to plant. This applies especially to the family of orchids, in which evolutionary processes are still recognizable. It does not need a very well trained nose to discover such differences, for example, among individuals of *Gymnadenia conopea*, a widespread terrestrial orchid of Europe. In a comparative study the scents of 4 flowering plants of the same habitat comprising about 50 individuals have been trapped between 9 a.m. and 11:30 a.m. As can be seen from Table I the variation in the scent composition is considerable and it appears even allowable - if one includes the olfactory evaluation of a series of additional plants - to divide this habitat in an "eugenol-rich" and an "eugenol-poor" type of *Gymnadenia conopea*.

Scent and Pollination Principle

To attract the specific pollinator, in most cases insects, the orchid flowers use all imaginable shapes, colors and, above all, scents. Indeed, the enormous diversity in the scent of orchids may be considered as a reflection of the equally extreme diversity in pollination principles found within this family.

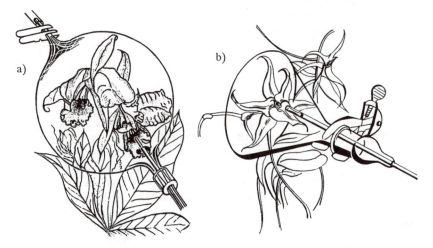

Figure 1.   Trapping the flower scent of a) *Cattleya labiata* b) *Angraecum sesquipedale*.

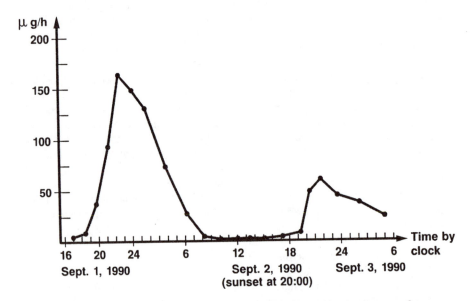

Figure 2.   *Epidendrum ciliare:* Emanation of scent in the course of two subsequent nights.

Table I. Variation in the Scent Composition on 4 Plants of *Gymnadenia conopea*

	1 %	2 %	3 %	4 %
Benzyl acetate	45.0	37.4	57.0	61.0
Benzyl benzoate	14.5	20.1	10.4	9.5
Methyl eugenol	5.7	6.2	5.9	6.1
Eugenol	2.6	9.5	0.3	0.2
Elemicine	3.3	3.1	5.2	5.4
Benzyl alcohol	2.6	2.5	1.4	1.5
(E)-Cinnamic alcohol	1.6	1.1	0.3	0.1
Phenylethyl alcohol	0.4	2.3	0.2	0.1
Phenylethyl acetate	0.4	0.5	1.4	1.3
(Z)-3-Hexenol	0.4	0.6	0.2	0.2

All samples trapped on charcoal (5mg) between 9 a.m. and 11:30 a.m.
Hostig, Uster, Switzerland, June 30, 1990.

"Moth Flowers". At the one extreme, there are the moth-pollinated, night-scented orchids, which are in most cases characterized by highly attractive so-called "white-floral" fragrances, fragrances reminiscent of e.g. jasmine, honeysuckle, tuberose, lily, gardenia and night-scented *Nicotiana* species. These night-active orchid flowers are not only of white color, they also show all the other aspects of the syndrome associated with moth pollination.

Angraecum Species. One of the most fascinating examples of this type of co-evolution/adaption between flower and pollinator is given by the Madagascar orchid *Angraecum sesquipedale*, which displays exclusively during the night its scent, reminiscent of lily and some night-active *Nicotiana* species. The species name "sesquipedale" refers to the spur which may measure up to "a foot and a half"(normally 25 - 30 cm) and it appears obvious that the pollinator has to be very long-tongued to get the nectar from the bottom of the spur.

As indicated in Table II, the scent of this spectacular orchid, which early attracted Darwin's interest, is rich in isovaleraldoxime ($\underline{3}$, E:Z ~ 2:1) and phenyl-acetaldoxime ($\underline{4}$) (see Figure 3). We have been able to identify such oximes in the floral scents of 14 orchid species and, remarkably, the majority of these species is characterized by the syndrome of moth pollination. These oximes $\underline{1}$ - $\underline{4}$, which are often accompanied by the corresponding nitriles and nitro compounds as minor components, may be regarded as being derived from the corresponding amino acids (*17,18*).

Another representative of this genus - *Angraecum eburneum* - is well suited to illustrate that trapping techniques, in combination with such characteristic constituents, might also be useful in the field of taxonomic investigation. Thus, the flower scent of *A. eburneum* ssp. *eburneum* is rich in isovaleraldoxime ($\underline{3}$, ~ 10%, trapped from 7 p.m. to 10 p.m.) and other derivatives of leucine, while 2-methylbutyraldoxime ($\underline{2}$, ~ 26%, trapped from 6 p.m. to 10 p.m.) is characteristic for the subspecies *superbum*.

Aerangis Species. Another extremely attractive African genus, which comprises practically only white-colored, night-scented species is that of *Aerangis*. Typical examples are *A. confusa* native to Kenya and *A. kirkii*, native to Kenya and Tansania, both species emitting an attractive "white-floral" scent characterized by aspects reminiscent of tuberose and gardenia. For the latter characteristic the same new natural product is responsible in both species, the *cis*-4-methyl-5-decanolide ($\underline{9}$) shown in Figure 4. Since we could identify this attractively scented lactone up to now only in these two *Aerangis* species, we would like to name it Aerangis lactone. In *A. confusa* it is present at 2 - 5 % and in *A. kirkii* at the extremely high level of 20 - 30 % (percentage values show pronounced time-dependence). The Aerangis lactone ($\underline{9}$) is easily accessible together with some of its trans-isomer $\underline{8}$ by hydrogenation of dihydrojasmone ($\underline{5}$) to the cyclopentanones $\underline{6}$ and $\underline{7}$, followed by Baeyer-Villiger oxidation. The absolute configuration of $\underline{9}$ has not yet been established.

In *A. kirkii*, this lactone $\underline{9}$ is accompanied to about 0.6 % by the *cis*-3-methyl-4-decanolide ($\underline{11}$) which is accessible together with the *trans*-isomer $\underline{10}$

Table II.   Composition of the Trapped Scent of *Angraecum sesquipedale*

Compound	Area %*
Isovaleraldehyde	2.50
Isovaleronitrile	3.50
Isoamyl acetate	0.50
Limonene	0.50
Isoamyl alcohol	0.90
(Z)-3-Hexenol	0.20
Isovaleraldoxime (E/Z ca. 2:1)	34.00
Benzaldehyde	1.60
Linalool	0.70
Methyl benzoate	17.90
Phenylacetaldehyde	0.30
Ethyl benzoate	0.30
Benzyl acetate	0.80
Hydroquinone dimethyl ether	0.40
Neral	0.20
Methyl salicylate	0.20
cis-Linalooloxide (pyranoid)	0.10
Geranial	0.20
Geranyl acetate	0.10
Geraniol	0.30
Benzyl butyrate	0.20
Benzyl isovalerate	0.20
Benzyl alcohol	14.80
Phenylethyl alcohol	2.50
ß-Ionone	0.30
ß-Ionone epoxide	0.05
Anisaldehyde	0.60
(Z)-3-Hexenyl benzoate	0.05
Anisyl acetate	0.05
Methyl anthranilate	0.20
Anisyl alcohol	0.10
(E)-Cinnamic alcohol	0.40
Dihydroactinidiolide	0.20
Phenylacetaldoxime	2.00
Indole	0.50
Benzyl benzoate	3.00
Phenylethyl benzoate	0.30
p-Methoxy cinnamic alcohol	0.20
Benzyl salicylate	0.60

*second night after inflorescence, 8 p.m. - 7 a.m., Feb. 1/2, 1988
relative and absolute amounts of volatiles show pronounced time-dependence

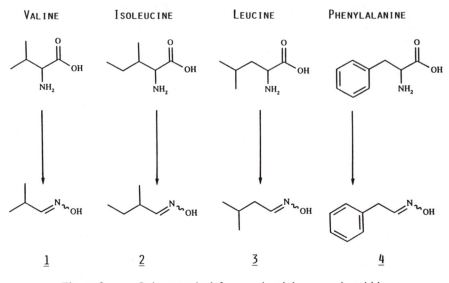

Figure 3.   Oximes typical for certain night-scented orchids.

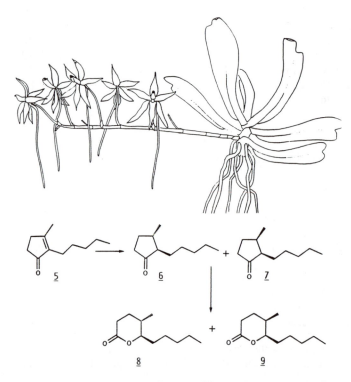

Figure 4.   Aerangis lactone (9), an important scent constituent of *Aerangis confusa* (shown as drawing) and *A. kirkii*.

(10:11 ~ 2:1) by application of the known radical addition of an alcohol to a crotonate, in this case n-heptanol to methyl crotonate (Figure 5). The *cis*-lactone 11 has not yet been described as occurring in nature while the *trans*-isomer 10 was found by Maurer & Hauser (*19*) as a trace constituent in clary sage oil. The lactone 11 is characterized by fruity-floral aspects reminiscent of peach and gardenia.

Furthermore, the scent of *A. confusa* contains around 30 % of the structurally interesting C_{16}-homoterpene 12, which - according to recent experimental confirmation by Boland (*20*) - has to be considered as a metabolite of geranyllinalool (Figure 6). This tetraene 12, which was described by Maurer and co-authors (*21*) as new natural product occurring in cardamom oil, has been identified in the course of this project in about 30 orchid species. Very recently Dicke and co-authors (*22*) discovered that 12, in combination with other volatiles, is released if leaves of the lima bean *Phaseolus lunatus* are damaged by the spider mite *Tetranychus urticae*.

Remarkably, in the scent of *A.confusa* as well as in that of *A.kotschyana*, the C_{16}-homoterpene 12 is accompanied by the methyl esters 13 - 15, which are possibly also derived from geranyllinalool. The new methyl esters 13 and 15, and α-farnesenic acid methyl ester (14), all easily accessible from (E)-geranylacetone, might be of taxonomic importance. As shown in Figure 7, a subspecies of *A.confusa,* occurring at different habitats in Kenya, shows a very similar pattern of volatiles as the common type except for these 3 esters which are totally missing. An overview on the composition of these two types of *A.confusa* is given in Table III.

The Scents of some Angraecoid Orchids in Comparison to those of *Brassavola* Species. These night-scented, white-colored, moth-pollinated orchids are highly characteristic of the African orchid flora, which hosts most likely over 1200 representatives of this ecological group. This number is based on the estimation, that

- around 10 % of all orchid species are native to Africa (*23*)
- that around 8 % of all orchids are characterized by the syndrome of
  moth pollination (*2*) and
- that over 60 % of the moth-pollinated orchids (own estimation) are native
  to Africa including Madagascar.

Most of them belong to genera of the tribe Vandeae as do the examples of the genera *Angraecum* and *Aerangis*, which have been discussed and which belong to the so-called group of angraecoid orchids.

In the neotropics "moth orchids" are much less represented and within the Indoaustralian distribution area, they occur only as very rare exceptions. The comparison of the basic scent composition of the angraecoid orchids investigated with that of the neotropic genus *Brassavola* reveals some characteristic differences, which might be of interest regarding evolutionary aspects of these genera.

The scents of the angraecoid orchids of the genera *Aerangis* and *Angraecum* (Figure 8) are often based on the acyclic tertiary terpene alcohols linalool and nerolidol and/or relatively simple aromatic compounds, especially benzyl and

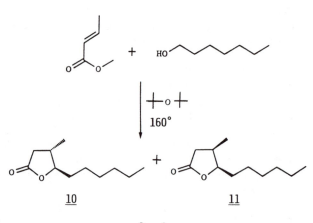

2 : 1

Figure 5.   *cis*-3-Methyl-4-decanolide (**11**), a minor scent constituent of *Aerangis kirkii.*

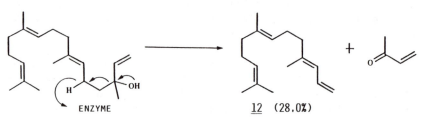

GERANYLLINALOOL

12 (28.0%)

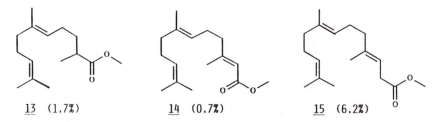

13 (1.7%)   **14** (0.7%)   **15** (6.2%)

Figure 6.   Interesting metabolites of geranyllinalool in the scent of *Aerangis* species. Percentage values refer to *Aerangis confusa* (6 p.m. - 8 p.m., Nov. 28, 1990).

phenylethyl alcohol including derivatives and esters of benzoic as well as salicylic acid. Quantitative differences within this common scent part and/or specific compounds, as those summarized for the examples in Figure 8, finally characterize the scents of the individual species. From six representatives of the neotropic genus *Brassavola* comprising about 15 species, three (*B.flagellaris, B.tuberculata, B.nodosa*) are also based on this principle, while the other three (*B.digbyana, B.glauca*, another clone of *B.nodosa)* contain additionally citronellol or geraniol as main component, accompanied on the level of minor components by corresponding derivatives. These two primary monoterpene alcohols are often found in the scents of day-active flowers with "rosy-floral" character and they are rather rare in those of night-scented species. In the scents of the 11 species of *Aerangis* and 7 species of *Angraecum* investigated so far, they occur at most in minor amounts. Furthermore, (E)-ocimene seems to occur in the scents of most *Brassavola* species as major or main component, while it was found only as minor or trace constituent in the angraecoid orchids investigated.

"Fly Flowers" and related Flower Types. In dramatic contrast to the night-scented species, which are, in most cases, also highly attractive to the human nose, some orchid flowers mimic carrion with their scent, texture and color. Carrion flies are attracted believing the flower would really be putrefied meat. They deposit their eggs on the labellum and the flower is consequently pollinated, the purpose of all this mimicry.

This phenomenon is especially well represented in the extensive *Cirrhopetalum* genus, which is spread across the whole of Southeast Asia. One unforgettable example is *C. robustum* from the rain forests of New Guinea, whose flowers emit a particularly penetrating stench containing butyric acid ($\sim 4\%$), 2-methylbutyric acid ($\sim 3\%$) and homologues. At the same time an aminic note reminiscent of N-methyl and N,N-dimethyl amine is recognizable. The compounds responsible for these olfactory aspects have not yet been identified.

The South American genus *Masdevallia* also comprises species with similar penetrating scents. A very typical example is *Masdevallia caesia,* which grows in the mountain cloud rain forests of Colombia at altitudes of 2000 - 2500 m. Its repulsive flower scent contains butyric acid ($\sim 3.5\%$) and isovaleric acid ($\sim 0.5\%$) and the unusual components phenylglyoxal ($\sim 20\%$) and α-hydroxy acetophenone ($\sim 8\%$).

Other species of this ecological group of orchids pollinated by flies, mosquitos and gnats, which represents in total about 15% of all species (2), emit scents reminiscent of seaweed, algae, crustacea, mushrooms etc.

Dracula chestertonii.   As an illustration of this incredible talent for mimicry, we shall briefly consider the extraordinarily shaped flower of *Dracula chestertonii* (Figure 9), an inhabitant of Colombia, which gives off a mushroom-like scent. According to Vogel (*24*) females of a particular fungus-fly are attracted by this semiochemical signal. Furthermore, the large labellum of this *Dracula* species imitates so deceptively in size and shape the head of mushrooms found in the same biotope, that the fungus-fly becomes a victim of this mimicry and deposits its eggs on the lip.

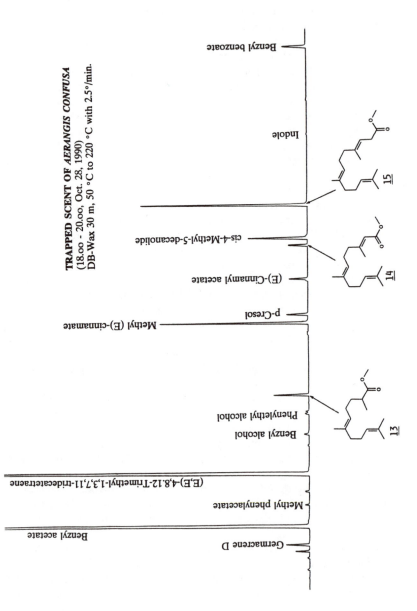

TRAPPED SCENT OF *AERANGIS CONFUSA*
(18.oo - 20.oo, Oct. 28, 1990)
DB-Wax 30 m, 50 °C to 220 °C with 2.5°/min.

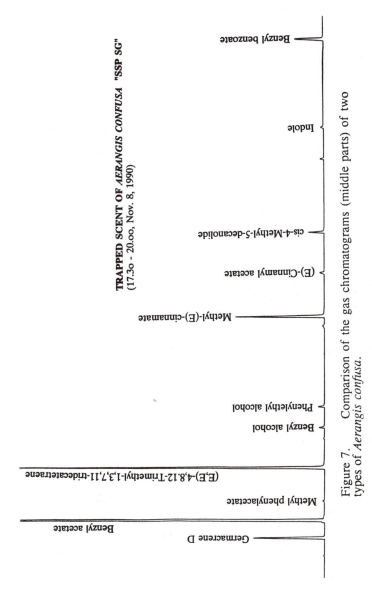

Figure 7. Comparison of the gas chromatograms (middle parts) of two types of *Aerangis confusa.*

Table III. Comparison of the Scent Composition of Two Types of *Aerangis confusa*

Compound	Area %[a]	Area %[b]
α-Pinene	0.30	0.10
Limonene	1.50	0.10
Eucalyptol	0.30	0.10
(E)-Ocimene	0.02	0.10
Nonanal	0.10	-
Methyl 3-methyloctanoate	3.20	1.10
Benzaldehyde	1.80	2.50
Linalool	2.20	1.50
Methyl benzoate	4.50	16.00
Benzyl formate	0.50	0.10
Germacrene D	0.80	5.40
Benzyl acetate	31.20	31.00
Methyl phenylacetate	0.10	0.20
(Z,E)-4,8,12-Trimethyl-1,3,7,11-tridecatetraene	0.10	0.10
(E,E)-4,8,12-Trimethyl-1,3,7,11-tridecatetraene (12)	28.00	26.00
Phenylethyl acetate	1.10	-
Benzyl alcohol	0.30	0.40
Phenylethyl alcohol	0.10	0.40
Methyl (E)-2,6,10-trimethyl-5,9-undecadienoate (13)	1.30	-
(E)-Nerolidol	0.20	0.20
Methyl (E)-cinnamate	5.20	6.40
p-Cresol	0.70	0.10
(E)-Cinnamyl acetate	0.80	0.50
Methyl (E,E)-3,7,11-trimethyl-2,6,10-dodecatrienoate (14)	0.70	-
cis-4-Methyl-5-decanolide (9)	2.70	1.50
Methyl (E,E)-4,8,12-trimethyl-3,7,11-tridecatrienoate (15)	6.20	-
Indole	0.10	0.20
Benzyl benzoate	1.20	2.50
Benzyl salicylate	0.20	0.30

a) common type,   6 p.m. - 8 p.m., Oct. 28, 1990
b) "subspecies SG", 5:30 p.m. - 8 p.m., Nov. 8, 1990
relative and absolute amounts of volatiles show pronounced time-dependence

Aerangis kirkii [a)]

Methyl 3-methyloctanoate (2.0 %)
Germacrene D (33.0 %)
p-Cresol (1.0 %)
cis-3-Methyl-4-decanolide (0.6 %)
Aerangis lactone (26.2 %)

Aerangis brachycarpa [b)]

Limonene (8.2 %)
p-Methyl anisol (4.3 %)
Anethole (6.6 %)
p-Cresol (0.4 %)
Indole (2.3 %)

Linalool, Nerolidol
Benzyl/Phenylethyl alcohol and Esters
Esters of Benzoic/Salicylic acid

Isovaleraldoxime (0.4 %)
Phenylacetaldoxime (0.3 %)
Anethole (0.4 %)
Eugenol (1.8 %)
(E)-Isoeugenol (0.3 %)

Isovaleraldoxime (9.9 %)
Benzaldehyde (12.0 %)
Phenylacetaldehyde (3.0 %)
Phenylglyoxal (0.5 %)
Indole (1.0 %)

Angraecum bosseri [c)]

Angraecum eburneum [d)]

Figure 8.   Basic scent composition of angraecoid orchids of the genera *Aerangis* and *Angraecum*. a) 7:30 p.m. - 11 p.m., April 15,1991. b) 8 p.m. - 11 p.m., Oct. 26,1990. c) 7 p.m. - 10 p.m., Dec. 5, 1990. d) ssp.*eburneum*, 7 p.m. - 10 p.m., Jan.7, 1991.

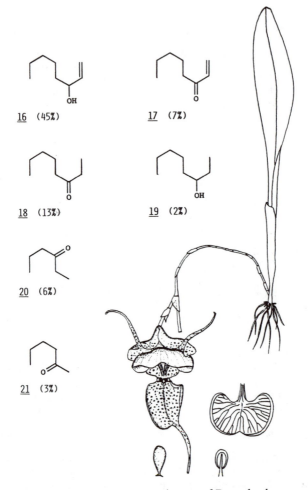

16   (45%)

17   (7%)

18   (13%)

19   (2%)

20   (6%)

21   (3%)

Figure 9.   Characteristic scent constituents of *Dracula chestertonii* (shown as drawing).

Within this project we had the opportunity to identify some components possibly important to this extraordinary example of mimicry. Thus, the scent constituents 16 - 21 summarized in Figure 9 could be identified, all of them known to occur also in mushrooms. Especially typical for the characteristic scent are   1-octen-3-ol (16) and its derivatives 17 - 19. As experienced during this investigation, the intensity of scent emitted by these unique flowers may vary considerably from clone to clone and in the best case, not more than 1 - 2 μg of volatiles could be accumulated from a flower during the trapping time of two hours (noon to 2 p.m.). Therefore, the given percentage values should only be taken as rough estimations.

"Bee Flowers" and related Flower Types. Between these two extreme cases of ecological groups are thousands of mainly day-scented, tropical and subtropical orchids attracting with all imaginable floral fragrances the respective bee, bumblebee or wasp species. These fragrances cover the entire range of "rosy-floral", "ionone-floral", "spicy-floral" and "animalic-floral" notes including all possible combinations thereof. Occasionally, these scents are very elegantly completed by woody, amberlike or musky notes, or are even characterized by one of these olfactory aspects as it will be discussed in an example. According to van der Pijl & Dodson (2) around 60 % of all orchids are pollinated by such Hymenoptera species and, therefore, only a very few special examples may be discussed in this context.

Cattleya **Species.** The appearance of a *Cattleya* flower (compare Figure 1) is very typical for "bee flowers" among *Orchidaceae*. The bright colors of the three sepals and the two petals as well as of the especially intense lip support the attraction of the scent and the big platform-like lip invites the bee to land. A very characteristic example is the best known representative of this genus comprising about 60 neotropical species, *Cattleya labiata* from Brasil. The spectacular purple flowers emit a quite attractive aromatic "spicy-floral" fragrance (compare Figure 10), which is based on the olfactorily important major components linalool, methyl benzoate, methyl salicylate, phenylethyl alcohol, caryophyllene epoxide and eugenol and which is finally completed by a great variety of scent-characterizing minor and trace constituents. Major components not covered by the GC of Figure 10 are α-pinene (3.5%), limonene (8.3 %), (E)-ocimene (1.0 %), benzyl benzoate (0.8 %), phenylethyl benzoate (0.3 %) and benzyl salicylate (0.4 %). The olfactory characteristics of the scent of *Cattleya labiata* are recognizable in many representatives of the so-called group of "unifoliate Cattleyas".

An impressive olfactory exception within this group is given by *Cattleya percivaliana*, the national flower of Venezuela, whose scent is characterized by a dissonance reminiscent of the defense secretion of certain bedbugs. Responsible for this special note are unsaturated C_{10}-lipid metabolites shown in Figure 11, especially the (E,Z)-2,4-decadienal (22) and its (E,E)-isomer, which are accompanied by the methyl (E,Z)-2,4-decadienoate (23), the (E,Z)-2,4-decadienyl acetate (24) and related compounds such as the methyl (Z)-4-decenoate (25). The percentage values in Figure 11 marked by asterisks refer to the relative amounts of

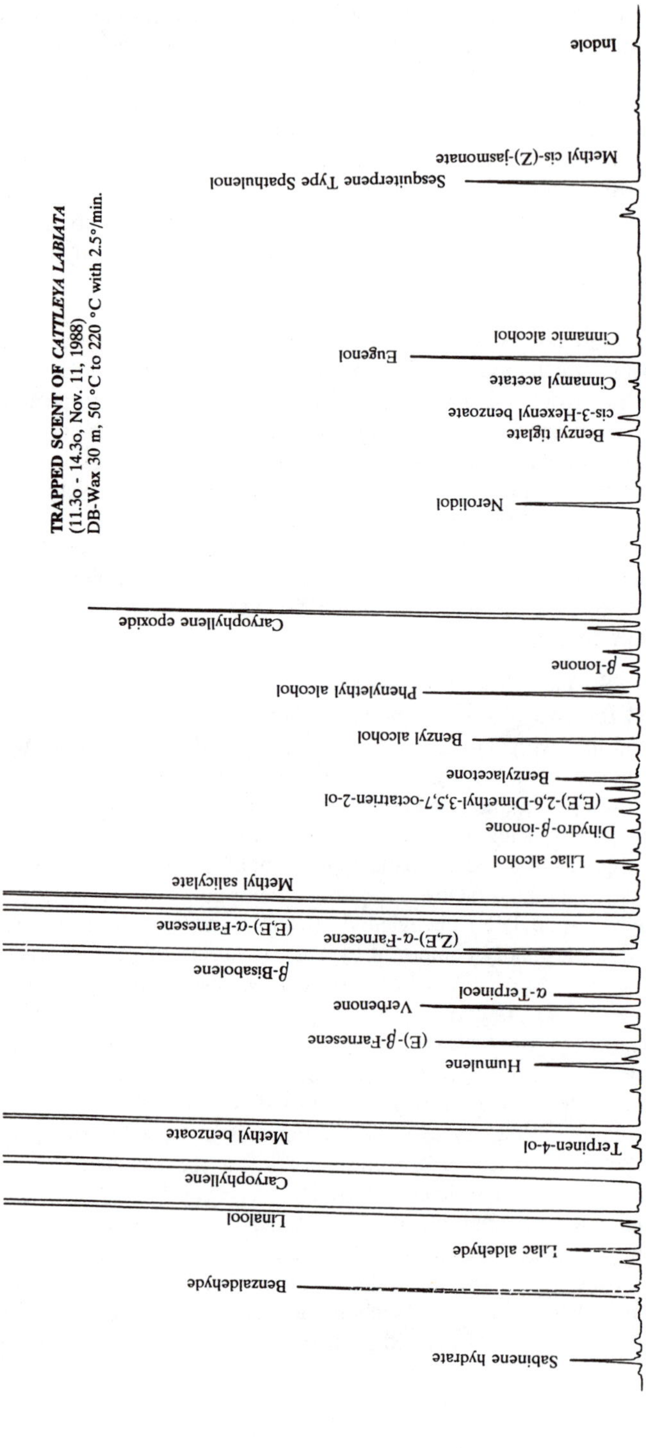

Figure 10. Gas chromatogram (middle part) of the trapped scent of *Cattleya labiata*, (11.30 a.m. - 2:30 p.m., Nov. 11, 1988).

the (E,E)-isomers. This peculiar note arising from the decadienals and derivatives is perceptible in the scents of about 80 species out of the 2200 species olfactorily evaluated. The astonishing extreme is given by the intense and penetrating scent of *Rodriguezia refracta* native to Ecuador, which consists practically only of (E,Z)-2,4-decadienal (22).

A second highly interesting exception within the "unifoliate Cattleyas" is *C. luteola*, which is native to the Brazilian states of Amazonas and Pará and adjacent regions of Peru and which is generally considered as being scentless. However, if one evaluates its elegant yellow flowers in the early morning between 4 and 6 o'clock, a quite delicate fresh, green-floral, and slightly woody fragrance of moderate intensity is perceivable, which is hardly recognizable anymore after 7 a.m. This applies especially to the pure yellow type of *C. luteola*, which does not have red stripes or spots on the inside of the lateral lobes of the labellum. Flowers with red on the labellum seem to emit their fragrance, which is of slightly different quality and lower intensity, over a longer period, from about 2 a.m. until 10 a.m. The scent of the pure yellow type investigated also by analytical means consists of over 95% of sesquiterpene hydrocarbons of rather high threshold values such as caryophyllene and α-copaene and only about 1% of potent scent donating and characterizing constituents (Table IV).

Cattleya luteola illustrates not only that the scent emanation of certain species may take place during a short period of day or night but also, that the intensity of the scent - as perceived by the human nose - may not be a measure for the absolute amount of volatiles emitted by the respective species. As observed by Dodson (2) during field work in Peru, *C. luteola* is visited and pollinated between 5.3o a.m. and 5.45 a.m. by crepuscular bee species and the emission of the fragrance seems to be optimally synchronized with the short visiting period of the pollinator.

The (E,E)-2,6-dimethyl-3,5,7-octatrien-2-ol and the (E,E)-2,6-dimethyl-1,3,5,7-octatetraene listed in Table IV as well as the corresponding (E,Z)-isomers have been described by Kaiser & Lamparsky (25) 15 years ago as occurring in the scent of hyacinth and more recently by Kaiser (12) as constituents of many additional flower scents. According to very recent findings of Brunke, Hammerschmidt & Schmaus (26), however, these compounds have to be considered as artefacts of ocimene, which are formed on the special quality of charcoal used for the trapping experiments. Depending on sample size and time of exposure to this adsorbent, these artefacts may represent from about 5% to over 50% of the original content of ocimene.

Masdevallia laucheana. A similar phenomenon of extreme time-dependence may be observed when following the scent emanation of *Masdevallia laucheana* from Costa Rica, which is also generally considered as being scentless. This is true for the day and night periods, however, during the 30 - 40 minutes of dusk, the characteristically shaped flowers emit a highly diffusive and most enjoyable fragrance characterized by "rosy-floral", "ionone-floral" and hesperidian-like aspects. As learnt from the sample accumulated during this short period of scent emanation, it is based on the main and major constituents (E)-

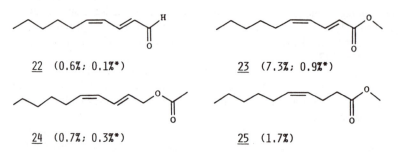

22 (0.6%; 0.1%*)   23 (7.3%; 0.9%*)

24 (0.7%; 0.3%*)   25 (1.7%)

Figure 11.   Characteristic scent constituents of *Cattleya percivaliana*.

Table IV. Composition of the Trapped Scent of *Cattleya luteola*

Compound	Area % *
α-Pinene	0.01
ß-Pinene	0.01
Myrcene	0.02
Limonene	0.70
(E)-Ocimene	0.30
Hexanol	0.02
(Z)-3-Hexenol	0.01
Nonanal	0.05
(E,E)-2,6-Dimethyl-1,3,5,7-octatetraene	0.30
Decanal	0.05
α-Copaene	4.50
Caryophyllene	83.50
Humulene	3.00
Germacrene D	0.30
α-Selinene	0.50
Phenylethyl acetate	0.02
(E,E)-4,8,12-Trimethyl-1,3,7,11-tridecatetraene	0.30
(E,E)-2,6-Dimethyl-3,5,7-octatrien-2-ol	0.40
Phenylethyl alcohol	0.05
Jasmone	0.02
Caryophyllene epoxide	0.30
Nerolidol	0.02
Benzyl benzoate	0.10

* trapped between 5 a.m. - 7 a.m. on Dec. 15, 1990;
  absolute amounts of volatiles show extreme time-dependence

ocimene (60.0%), neral (1.0%), geranial (6.0%), citronellol (2.5%), geraniol (20.6%), phenylacetonitrile (0.9%) and ß-ionone (0.5%). Additionally, this orchid scent contains to about 1.3% the new 3-oxo-7(E)-megastigmen-9-one (27), which is accompanied by trace amounts of the corresponding hydroxy ketone 29 (Figure 12). The diketone 27 together with the two hydroxy ketones 28 and 29 are accessible in a ratio of about 3:2:1 by application of the reaction sequence summarized in Figure 12 to the well known keto ester 26 (27). These carotenoid metabolites 27 - 29 could also be identified by us in the flower absolute of *Boronia megastigma* (together ca. 1%, 27:28:29 ~ 1:7:1) (28).

As a potential shorter synthesis of 27 - 29, the 3-oxo-α-ionol (30) was hydrogenated in the presence of Raney nickel and the hydroxy ketone thus obtained oxidized. However, this approach leads to the wrong series of isomers having an axial H_3C-C(5).

The olfactory differences between *Masdevallia laucheana* and *M. caesia*, which has already been briefly discussed as a species attracting carrion flies, could not be more extreme. In fact, the genus *Masdevallia* is a good example to illustrate that the enormous diversity in scents and pollination principles within *Orchidaceae* is sometimes even reflected within an individual genus. As another extreme, the vivid orange-scarlet colored flowers of *M. veitchiana,* occurring in Peru at elevations of 2200 - 4000 m, are practically scentless. It is generally assumed that hummingbirds which are mainly attracted by such vivid colors and do not have a well developed sense of smell (2) are the main-pollinators.

Bollea coelestis. One of the most exceptional scents among "bee-orchids" is emitted by *Bollea coelestis* native to Colombia, a spectacular representative of the so-called "fan-type orchids". However, it is again important to visit this orchid at the right time to experience its highly characteristic woody scent. During the first two to three days after anthesis, the flowers appear practically odorless, then they start to emit their rather intense scent. Two days later, the intensity has already decreased considerably, although the flowers may last for an additional week. This unusual flower scent is extremely rich in caryophyllene (33, 47%) and in a new sesquiterpene alcohol of molecular weight 222, which proved to be the caryophyll-5-en-2-α-ol (35, 42.3%) (Figure 13). The same scent contains a series of further sesquiterpene compounds such as humulene (1.2%), germacrene A (3.1%), caryophyllen ß-epoxide (1.0%), caryophyllene epoxide (4.3%) and humulene epoxide II (0.1%). The new tertiary alcohol 35 is formed to about 0.5%, together with other side products, by the oxymercuration/ cyclization/reduction of caryophyllene (33) to the caryophyllan-2,6-α-oxide (34), which we synthesized some years ago in connection with the structural elucidation of the corresponding ß-oxide found in *Lippia citriodora* (29). Selectively, it is accessible from the described diepoxide 37 (30,31) by reductive cleavage of the 2(12)-epoxy group with lithium aluminum hydride to the epoxy alcohol 38 followed by desepoxidation with Zn/Cu-couple (29). The flower scent of *Bollea coelestis* is strongly dominated by the woody odor of this structurally interesting caryophyllene derivative 35 (see Table V).

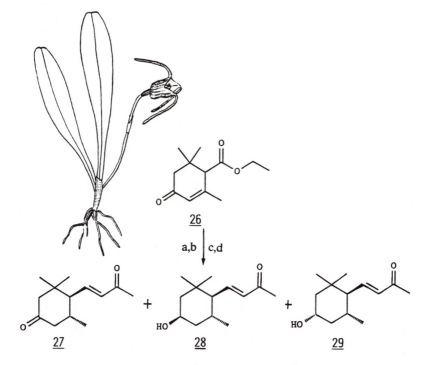

a) LiAlH₄,THF   b) H₂,PtO₂,EtOH   c) PCC,CH₂Cl₂   d) acetone/OH⁻

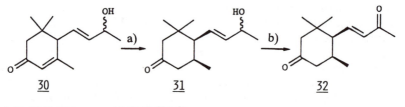

a) H₂,Ni,EtOH   b) PCC,CH₂Cl₂

Figure 12.   New carotenoid metabolites (40 and 42) in the scent of
Masdevallia laucheana.

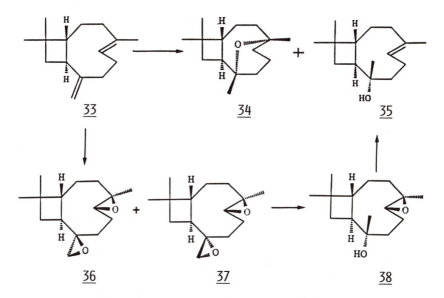

Figure 13. Caryophyll-5-en-2-α-ol (48), a main constituent of the scent of *Bollea coelestis.*

Cymbidium goeringii. To conclude the ecological group of "bee orchids" and to end at the same time this brief overview on the scent of orchids, no other subject is better suited than the discussion of one of the famous "to-yo-ran" orchids which have for centuries enjoyed a high cultural image in China and Japan (*32,33*). Especially attractive in this context appears to be *Cymbidium goeringii*, which is native to southern China, Formosa and southern Japan, and which occurs in many varieties. The widespread Chinese type, which we investigated, is characterized by a beautiful clean and transparent scent reminiscent of fully ripe lemon fruits, lily of the valley and methyl jasmonate. Indeed, the trapped flower scent contains not only the methyl *cis*-(Z)-jasmonate (40) accompanied by the *trans*-(Z)-isomer 39 but even to a much higher amount the new methyl *cis*-(Z)-dehydrojasmonate (45) and the *trans*-(Z)-isomer (43) (Figure 14), which are the main contributors to the unique scent of *C. goeringii*. For the clean "lily of the valley"-type aspect, the main constituents nerolidol (58.0%), (E,E)-farnesol (11.5%) and (E,E)-farnesal (1.5%) and their derivatives are responsible.

For the structural proof of the new jasmonoids 43 and 45, commercially available methyl jasmonate (39/40 ~ 88 : 12) was ketalized with ethylene glycol and subsequently transformed via ozonolysis to the aldehyde 41. Finally, Wittig reaction of 41 with allyltriphenyl phosphorane followed by deketalization gave a mixture of the methyl dehydrojasmonates 42 - 45 in a ratio of 46 : 44 : 6 : 4. The isomers 42 and 43 could be isolated as pure compounds by preparative GC, while the methyl cis-(Z)-dehydrojasmonate 45 was only obtained together with the cis-(E)-isomer 44. Equilibration of this 3 : 2 mixture of 44 and 45 with sodium methoxide in methanol gave the corresponding trans-isomers 42 and 43 in the same ratio.

Table V.   Spectral Data of <u>9</u>, <u>11</u>, <u>13</u>, <u>15</u>, <u>27</u>, <u>28</u>, <u>29</u>, <u>35</u>, <u>43</u>, <u>45</u>

<u>9</u>.   -MS.: 184 (M$^+$, 0.2), 128(3), 113(23), 99(3), 85(15), 84(39), 69(5), 56(100), 55(24), 43(26), 41(22). -^{1}H-NMR (400 MHz).: 0.90(t,J~7, 3H); 0.96(d,J~7, 3H); 1.25-1.60(m,together 6H); 1.62-1.72(m,2H); 1.95-2.10(m, 3H); 2.53 (dxd, J~7, 2H); 4.28(m,1H). -^{13}C-NMR.:CH$_3$-signals at 12.60(C(11)); 14.04 (C(10)). CH$_2$-signals at 22.58; 25.32; 26.21; 26.85; 31.70; 32.01. CH-signals at 29.44(C(4)); 83.07(C(5)). C-signals at 171.96(C(1)). -IR.: 1735, 1238, 1200, 1140, 1095, 1069, 1054, 993, 908.

<u>11</u>.   -MS.: 184(M$^+$, 0.03), 142(11), 124(7), 115(10), 99(100), 97(18), 71(32), 55(31), 43(37), 41(24). -^{1}H-NMR(200MHz).: 0.90(t,J~7, 3H); 1.02(d,J~7, 3H); 1.22-1.40(m,8H), 1.40-1.75(m,2H) 2.20(m,1H); 2.50-2.76(m,2H); 4.42 (m,1H). (cis-configuration was established by NOE-measurements). -IR.: 1780, 1385, 1340, 1295, 1212, 1170, 972, 934.

<u>13</u>.   -MS.: 236(M$^+$,1), 195(28), 163(19), 137(10), 123(11), 109(92), 95(14), 88(25), 81(27), 69(100), 67(35), 59(8), 55(18), 41(81).

<u>15</u>.   -MS.: 264(M$^+$,2), 221(2), 180(2), 153(3), 136(23), 121(36), 107(7), 93(16), 85(12), 81(35), 69(100), 59(6), 55(7), 41(48).

<u>27</u>.   -MS.: 208(M$^+$,5), 190(3), 152(2), 147(8), 124(5), 111(17), 110(41), 109(17), 95(100), 83(9), 81(11), 69(5), 67(9), 55(9), 43(43), 41(18). -^{1}H-NMR (400MHz).: 0.90(s,3H); 0.94(d,J~6, 3H); 0.98(s,3H); 2.02 - 2.08(H$_{ax}$-C (4), H$_{ax}$-C(5) and H$_{ax}$-C(6)); 2.16(H$_A$-C(2),J$_{AB}$~13.5, J$_{AX}$~2); 2.32(H$_B$-C(2), J$_{AB}$ ~13.5); 2.44(m,H$_{eq}$-C(4)); 6.18(d,J~16,H-C(8)); 6.56(dxd,J$_1$~16,J$_2$~9, H-C (7)). -IR(CHCl$_3$).: 1700, 1665, 1620, 985.

<u>28</u>.   -MS.: 210(M$^+$,1), 177(3), 154(7), 149(22), 136(6), 121(8), 111(79), 109(20), 95(81), 93(29), 84(26), 81(20), 69(14), 67(19), 57(17), 55(18), 43(100), 41(42).- ^{1}H-NMR(400MHz).: 0.80(d,J~6.5, H$_3$C-C(5)); 0.82(s,3H); 1.12 (s, 3H); 1.23(m, H$_{ax}$-C(4)); 1.42(dxd, J$_1$~14, J$_2$~3, H$_{ax}$-C(2)); 1.57(t,J~10, H$_{ax}$-C(6)); 1.66(dxt, J$_1$~14, J$_2$~2, H$_{eq}$-C(2)); 1.85(dxq, J$_1$~14, J$_2$~2, H$_{eq}$-C(4)); 2.03(m,H$_{ax}$-C(5)); 2.28(s,3H); 4.16(bs,H$_{eq}$-C(3)); 6.05(d,J~16, H-C(8)); 6.62 (dxd, J$_1$~16, J$_2$~10, H-C(7)). -IR (CHCl$_3$).: 3600, 3450, 1670, 1620, 995. (b = broad).

Table V. continued

---

29. -MS.: 210(M$^+$,1), 192(1), 177(3), 154(10), 149(30), 136(8), 121(9), 111(100), 109(19), 95(95), 93(35), 84(30), 82(24), 69(12), 67(23), 57(21), 55(18), 43 (99), 41(39).-^{1}H-NMR(400MHz).: 0.83(d,J ~ 6.5, H$_3$C-C(5)); 0.89(s,3H); 0.92 (s,3H); 0.96(q,J ~ 11.5, H$_{ax}$-C(4)); 1.16(t,J ~ 11.5, H$_{ax}$-C(2)); 1.52(t,J ~ 10, H$_{ax}$-C(6)); 1.68(m,H$_{ax}$-C(5)); 1.78(dxdxd, J$_1$ ~ 12.5, J$_2$ ~ 4, J$_3$ ~ 2, H$_{eq}$-C(2)); 2.06(m, H$_{eq}$-C(4)); 2.26(s,3H); 3.83(txt, J$_1$ ~ 11.5, J$_2$ ~ 7, H$_{ax}$-C(3)); 6.08 (d, J ~ 16, H-C (8)); 6.52(dxd,J$_1$ ~ 16, J$_2$ ~ 10, H-C(7)).- IR(CHCl$_3$).: 3590,3430,1660,1615,985.

35. -MS.: 222(M$^+$,0.5), 166(5), 161(6), 151(47), 148(23), 123(15), 109(35), 108(36), 95(38), 93(57), 81(69), 71(33), 69(39), 67(35), 59(30), 55(35), 43 (100), 41(65). -^{1}H-NMR(400MHz,CDCl$_3$,20°C).: 0.95(s,2xCH$_3$-C(10) and CH$_3$-C(2)); 1.66(bs,CH$_3$-C(6)); 1.20-2.30(m, broad signals,13H); 5.18-5.40 (bm,H-C(5)). In CDCl$_3$ at 20°C, some signals of the ^{13}C-NMR appear extremely broad and are not interpretable anymore. As a compromise between quality of signals and dehydration/isomerization of 35 at elevated temperature, spectra have been measured in DMSO at 95°C. -^{1}H-NMR (400MHz, DMSO,95°C).: 0.94 and 0.95(2xs,2xCH$_3$-C(10)); 1.08(s,CH$_3$-C(2)); 1.63(s,CH$_3$-C(6)); 1.20-2.30(m,13H); 5.22(m,H-C(5)). -^{13}C-NMR (DMSO,95°C).: CH$_3$-signals at 16.68(CH$_3$-C(6)); 19.38(CH$_3$-C(2)); 22.73+ 22.78(2xCH$_3$-C(10)); CH$_2$-signals at 22.34(C(4)); 28.91(C(7)); 37.66; 43.40; CH-signals at 47.74(C(9) or C(1)); 51.83(C(9) or C(1)); 123.09 (C(5)); C-signal at 30.30. -IR(CHCl$_3$).: 3590, 3430, 1090, 1055, 950, 875.

43. -MS.: 222(M$^+$,13), 204(11), 193(11), 149(79), 144(41), 131(55), 130(79), 107(73), 105(59), 91(81), 79(100), 67(98), 59(26), 55(58), 41(85). -^{1}H-NMR (400MHz).: 1.51(m,1H); 1.94(m,1H); 2.12(m,1H); 2.20-2.42(m,4H); 2.51 (m,2H); 2.66(m,1H); 3.70(s,3H); 5.15(d,J ~ 10.5, 1H); 5.22(dxd,J$_1$ ~ 1.5, J$_2$ ~ 17, 1H); 5.39(dxt,J$_1$ ~ 8, J$_2$ ~ 10.5, 1H); 6.08(t,J ~ 10.5, 1H); 6.63(dxtxd,J$_1$ ~ 1.5, J$_2$ ~ 10.5, J$_3$ ~ 17, 1H).

45. -MS.: 222(M$^+$,17), 204(22), 193(10), 149(30), 144(32), 131(53), 130(91), 129(35), 107(62), 91(74), 79(100), 67(98), 59(24), 55(52), 41(89). -^{1}H-NMR (400MHz).: 5.15(d,J ~ 10.5, 1H); 5.23(dxd,J$_1$ ~ 1.5, J$_2$ ~ 17, 1H); 5.43(dxt,J$_1$ ~ 8, J$_2$ ~ 10.5, 1H); 6.06(t,J ~ 10.5, 1H); 6.60(dxtxd,J$_1$ ~ 1.5, J$_2$ ~ 10.5, J$_3$ ~ 17, 1H).

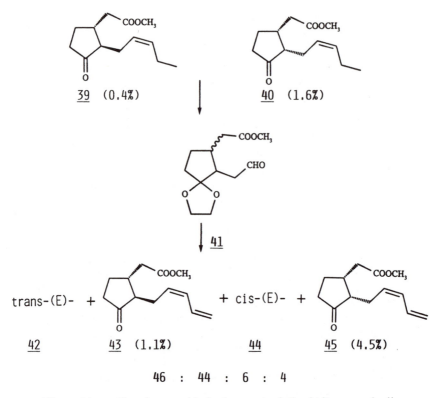

Figure 14.   New jasmonoids in the scent of *Cymbidium goeringii*.

Acknowledgments

I am grateful to Dr. J. Schmid for the GC/MS-measurements and many discussions on the spectra of new compounds; to Dr. E. Billeter and Mr. J. Märki for NMR investigations; to Dr. C. Nussbaumer for the synthesis of the caryophyll-5-en-2-α-ol (35) and the methyl dehydrojasmonates 43 and 45; to Mrs. A. Reolon-Huber for the skillful assistance throughout the project summarized in this paper; to Dr. Ch. Weymuth, University of Zurich, for a sample of the trapped scent of *Rodriguezia refracta* and discussions and Mr. H.P. Schumacher for the drawings in the Figures 4, 9 and 12. Last but not least, I have to express my sincere thanks once again to Mr. H.P. Schumacher, Botanical Garden of the City of St. Gallen, to Mr. W. Philipp, Botanical Garden of the University of Zurich and to Dr. K.H. Senghas, Botanical Garden of the University of Heidelberg, for the generous permission to collect at their premises scent samples of the orchids discussed in this paper.

Literature Cited

1.   Darwin, C., *The Various Contrivances by which Orchids are Fertilized by Insects.* 2nd ed. D. Appleton, New York, 1877 (only seen as summary).

2. Pijl, L. van der, & Dodson, C.H., *Orchid Flowers: Their Pollination and Evolution,* Univ. Miami Press, Coral Gables, Fla., 1969, pp. 83 - 87 and 128.

3. Williams, N.H., The biology of orchids and euglossine bees. Chapter 4 in *Orchid Biology, Reviews and Perspectives,* II., ed. J. Arditti. Comstock Publishing Associates, Cornell University Press, Ithaca and London, 1982, pp. 120 - 171.

4. Williams, N.H., Floral fragrances as cues in animal behavior. Chapter 3 in *Handbook of Experimental Pollination Biology,* Scientific and Academic Editions, New York, Cincinnati, Toronto, London & Melbourne, 1983, pp. 50 - 72.

5. Williams, N.H. & Whitten, W.M., Orchid floral fragrances and male euglossine bees: methods and advances in the last sesquidecade. *Biol. Bull.,* **1983,** *164,* 355 - 395.

6. Lindquist, N., Battiste, M.A., Whitten, W.M., Williams, N.H. & Strekowski L., *trans*-Carvone oxide, a monoterpene epoxide from the fragrance of *Catasetum, Phytochemistry,* **1985,** *24,* 863 - 865.

7. Whitten, W.M., Williams, N.H., Armbruster, W.S., Battiste, M.A., Strekowski, L. & Lindquist, N., Carvone oxide: An example of convergent evolution in euglossine pollinated plants. *Systematic Botany,* **1986,** *11,* 222 - 228.

8. Whitten, W.M., Hills, H.G. & Williams, N.H., Occurrence of ipsdienol in floral fragrances, *Phytochemistry,* **1988,** *29,* 2759 - 2760.

9. Gerlach, G. & Schill, R., Fragrance analyses, an aid to taxonomic relationships of the genus *Coryanthes. Pl. Syst. Evol.,* **1989,** *168,* 159-165.

10. Gerlach, G. & Schill, R., Composition of orchid scents attracting euglossine bees. *Bot. Acta,* **1991,** *104,* 379 - 391.

11. Bicchi, C. & Joulain, D., Headspace gas chromatographic analysis of medicinal and aromatic plants and flowers. *Flavour and Fragrance Journal,* **1990,** *5,* 131 - 145.

12. Kaiser, R., Trapping, investigation and reconstitution of flower scents. Chapter 7 in *Perfumes: Art, Science and Technology,* ed. P.M. Müller & D. Lamparsky, Elsevier Applied Science, London & New York, 1991, pp. 213 - 250.

13. Dobson, H.E.M., Analysis of flower and pollen volatiles. In *Modern Methods of Plant Analysis, New Series, Volume 12, Essential Oils and Waxes,* Springer, Berlin & Heidelberg, 1991, pp. 231 - 151.

14. Borg-Karlson, Anna-K., Chemical and ethological studies of pollination in the genus *Ophrys. Phytochemistry,* **1990,** *29,* 1359 - 1387.

15. Grob, K., Organic substances in potable water and its precursor. *J. Chromatogr.,* **1973,** *84,* 255 - 327.

16. Grob, K. & Zürcher, F., Stripping of trace organic substances from water; equipment and procedure. *J. Chromatogr.,* **1976,** *117,* 285 - 294.

17. Kindl, H. & Underhill, E.W., Biosynthesis of mustard oil glucosides; N-hydroxyphenylalanine, a precursor of glucotropaeolin and a substrate for the enzymatic and non-enzymatic formation of phenylacetaldehyde oxime. *Phytochemistry,* **1968,** *7,* 745 - 756.

18. Matsuo, M., Kirkland, D.F. & Underhill, E.W., 1-Nitro-2-phenylethane, a possible intermediate in the biosynthesis of benzylglucosinolate, *Phytochemistry*, **1972**, *11*, 697 - 701.

19. Maurer, B. & Hauser, A., New constituents of clary sage oil. In *Proceedings of the 9th International Congress of Essential Oils, Singapore, 1983*, Book 3, pp. 69 - 76.

20. Boland, W. & Gäbler, A., Biosynthesis of homoterpenes in higher plants. *Helv. Chim. Acta*, **1989**, *72*, 247 - 253.

21. Maurer, B., Hauser, A. & Froidevaux, J.-C., (E)-4,8-Dimethyl-1,3,7-nonatriene and (E,E)-4,8,12-trimethyl-1,3,7,11-tridectatetraene, two unusual hydrocarbons from cardamom oil. *Tetrahedron Lett.* **1986**, *27*, 2111 - 2112.

22. Dicke, M., Van Beek, T.A., Posthumus, M.A., Ben Dom, N., Van Bokhoven, H. & De Groot, AE., Isolation and identification of volatile kairomone that affects acarine predatorprey interaction. *J. Chem. Ecol.* **1990**, *16*, 381 - 396.

23. Stewart, J., Die Aerangis-Verwandtschaft, *Die Orchidee*, **1982**, *33*, 48-57.

24. Vogel, S., Pilzmückenblumen als Pilzmimeten. *Flora*, **1978**, *167*, 329-389.

25. Kaiser, R. & Lamparsky, D., Nouveaux constituants de l'absolue de jacinthe et leur comportement olfactif. *Parf. Cosm. Arômes,* **1977**, no.*17*, 71 - 79.

26. Schmaus, G., Personal communication of March 23, 1992. See also: Brunke, E.-J., Hammerschmidt, F.-J. & Schmaus, G., Flower scent of traditional medicinal plants. This book.

27. Surmatis, J.D., Walser, A., Gibas, J. & Thommen, R., A study on the condensation of mesityl oxide with acetoacetic ester. *J. Org. Chem.,* **1970**, *35*, 1053 - 1056.

28. Kaiser, R., Olfactory and chemical characteristics of floral scents. Paper presented at the 196th National ACS Meeting, Symposium on Progress in Essential Oil Research, Los Angeles, CA., September 1988.

29. Kaiser, R. & Lamparsky, D., Caryophyllan-2,6-ß-oxide, a new sesquiterpenoid compound from the oil of *Lippia citriodora* Kunth. *Helv. Chim. Acta,* **1976**, *59*, 1803 - 1808.

30. Maurer, B. & Hauser A., New sesquiterpenoids from clary sage oil, *Helv. Chim. Acta,* **1983**, *66*, 2223 - 2235.

31. Srinivasan, V. & Warnhoff, E.W., Base-catalyzed intramolecular displacements on certain 1,2-epoxides. *Can. J. Chem.,* **1976**, *54*, 1372 - 1382 and ref. cited therein.

32. Nakamura, S., Tokuda, K. & Omato, A., Japan Prize Fragrance Competition. *Am. Orchid Soc. Bull.,* **1990**, *59*, 1030 - 1036.

33. Pagani, C., Perfect men and true friends; the orchid in Chinese culture. *Am. Orchid Soc. Bull.,* **1991**, *60*, 1177 - 1183.

RECEIVED August 19, 1992

Chapter 19

Volatile Constituents of Roses

Characterization of Cultivars Based on the Headspace Analysis of Living Flower Emissions

I. Flament, C. Debonneville, and A. Furrer

Research Laboratories, Firmenich SA, P.O.B. 239, CH–1211, Geneva 8, Switzerland

The present work is based on the comparative headspace analysis of twenty-seven varieties of living roses. A classification of the volatile constituents into six functional groups including the most abundant and characteristic chemicals is proposed: hydrocarbons, alcohols, aldehydes and ketones, esters, aromatic ethers and other miscellaneous functions. A presentation of the analytical results is given in the form of a stylized flower (exogram): this graphical illustration allows at one glance a symbolization and recognition of odoriferous flowers or any other scenting living plant material.

Nature, and in particular the plant kingdom, consists of a complex assembly of production units providing many thousands of odoriferous substances. For insects and humans, the flower is certainly the part of the plant which is the most attractive to the senses of sight and smell. Classical extraction by solvents or vapor distillation of petals inevitably results in the irreversible degradation of a considerable number of their constituents and the isolation of an adulterated perfume. Such concentrates are evidently used for perfume creation but their analysis is not representative for precise analytical work, as the composition is very different from that of the natural starting material. In contrast, dynamic headspace analysis of living flowers is a gentle and convenient technique which allows the enrichment of the most delicate "essences", and the separation and identification of their individual components. Used successfully as early as 1977 for the analysis of *jacinth (1)* or *gardenia (2)* volatiles, the method has undergone a tremendous expansion throughout the eighties and is now considered as a routine technique in almost all laboratories working on volatile constituents of odorants (3-9) (5) (6) (7) (8) (9).

It is normal that the "mystical and magic" rose, due to its perfectly divine perfume, was the first to become the object of numerous classical analyses, the purpose being to identify the most noble and subtle of its odoriferous components. The task was not straightforward, demonstrated by the fact that, at the turn of this century, only eight constitutionally defined substances had been discovered in rose essence. In 1953 only 20 compounds were known and in 1961 the number had increased to 50. From the start of the sixties, modern analytical techniques resulted in an acceleration in the identification of the volatile components: 200 by 1970 and 400 by 1990. One of the most complete analyses was published by Kovats in 1987 *(10)* but the story of rose analysis is marked by milestones corresponding to the identification of minor characteristic volatile constituents. For example, the

NOTE: This chapter is dedicated to the memory of Dr. Alan F. Thomas.

0097–6156/93/0525–0269$06.00/0

Table I. Main constituents of some rose varieties

Cultivar®	(1)	(2)	(3)	(4)	(5)	(6)	(7)	(8)	(9)	(10)	(11)
SONIA MEILLAND	43.1			20.9							
MONICA		22.3	21.5								
Rosa rugosa rubra				53.9							
Rosa gallica				41.8							
Rosa muscosa purpurea				29.9	20.6						
CHATELAINE de LULLIER				26.1							23.0
Rose à parfum de l'Hay				23.2							
MARGARET MERRIL						39.5					
CONCORDE 92								48.8			
BARONNE de ROTHSCHILD								45.1	32.9		
Rosa damascena								29.7			
YONINA								26.5	26.2	29.0	
SUTTERS GOLD								23.7	28.8		
COCKTAIL 80								59.8			
SUSAN ANN				22.1				42.0			
ELISABETH of GLAMIS				19.4				24.5	20.4		
NUAGE PARFUME								19.3			
HIDALGO									43.2		
SYLVIA											67.3
YOUKI SAN											55.8
GINA LOLLOBRIGIDA											52.5
PAPA MEILLAND						29.4					40.3
WHITE SUCCESS											29.6

(1) Germacrene-D; (2) (*E*)-4,8-Dimethyl-1,3-7-nonatriene; (3) β-Caryophyllene;
(4) Phenylethanol; (5) Geraniol; (6) Citronellol; (7) Hexanol; (8) (Z)-3-Hexenyl acetate;
(9) Phenylethyl acetate; (10) Hexyl acetate; (11) Orcinol dimethylether.

identification of carvone and eugenol methylether in 1949 *(11)*, the structural elucidation and synthesis of rose oxide diastereoisomers in 1959-61 *(12,13)*, the discovery of rose furan in 1968 *(14)* and of 1-*p*-menthen-9-al in 1969 *(15)* and, last but not least, the isolation of ionones, dihydro-ionones, β-damascone and β-damascenone in 1970 *(16)*. The importance of such minor components in flavors and fragrances has been highlighted by Ohloff *(17)*.

The present work is based on the headspace analysis of 27 varieties of roses, including classical roses and more recent hybrids (**Trademark names® are capitalized**). The analysis was performed by a technique previously described *(18)*: adsorption on TENAX cartridges, subsequent thermal desorption, separation of constituents by gas chromatography on a wide-bore glass capillary column and identification by mass spectrometry (Finnigan MAT, ITD Mod. 800). Quantification was achieved by integration of the TIC (Total Ionization Current) diagrams. It has been shown by Kaiser *(19)* that the emission of volatiles in flowers is dramatically time dependent and that rhythmic changes of fragrance composition are observed during the course of diurnal cycles. On the other hand, Mookherjee *(3, 20)* has pinpointed the chemical differences in the smell between "living" and "dead" flowers: the composition of a picked tea rose is remarkably different from that of the corresponding living rose. For instance, (Z)-3-hexenyl acetate which constitutes 20% of the living headspace volatiles, is drastically reduced to 5% in the picked flower, whereas the concentration of 3,5-dimethoxytoluene (orcinol dimethylether) is doubled. As will be shown below we have corroborated these observations, but the changes are strongly dependent on the species and are not always so important. To ensure reproducibility in measurements, sampling of headspace volatiles was systematically performed on living flowers with the same type of cartridge, during the same period of time (8-9 a.m. to 4-5 p.m.) and under an identical air flow.

Table II.   Distribution of functionalities in the headspace constituents
of 27 cultivars

Cultivar®	R-H	R-OH	-COOR	Ar-O-
MONICA	**56.3**	1.8	5.8	13.8
SONIA MEILLAND	**46.1**	21.1	7.2	2.3
CHARLES de GAULLE	**40.7**	3.4	21.2	1.7
Rosa rugosa rubra		**81.0**	3.8	0.6
Rosa muscosa purpurea	9.6	**70.5**	11.2	0.4
MARGARET MERRIL	7.1	**62.5**	6.4	13.7
Rose à parfum de l'Hay	2.8	**56.5**	23.1	0.1
Rosa gallica	11.1	**47.5**	28.1	
WESTERLAND	0.1	**45.7**	29.3	2.9
Rosa damascena	0.4	**37.9**	45.5	0.8
CHATELAINE de LULLIER	0.2	**37.6**	23.0	23.1
BARONNE E. de ROTHSCHILD	1.8		**91.1**	0.5
SUTTERS GOLD	2.4	7.9	**75.5**	14.4
CONCORDE 92	2.4	4.1	**73.2**	13.9
COCKTAIL 80	1.0	7.7	**70.9**	16.3
SUSAN ANN	0.4	24.4	**59.7**	3.6
HIDALGO	2.5	2.4	**58.1**	29.7
YONINA	1.1	5.7	**55.8**	30.2
NUAGE PARFUME	4.5	21.1	**48.8**	17.3
ELISABETH of GLAMIS	7.1	20.4	**48.6**	6.1
JARDIN de BAGATELLE	22.8	0.2	**34.2**	10.8
PRELUDE	7.3	29.2	**30.9**	17.4
SYLVIA	11.6	0.3	0.7	**67.3**
YOUKI SAN	5.2		5.0	**58.8**
GINA LOLLOBRIGIDA	5.4	11.5	18.6	**53.7**
PAPA MEILLAND	1.4	31.5	16.2	**44.8**
WHITE SUCCESS	14.1	3.7	20.0	**30.1**

Headspace analytical results are usually presented as lists of compounds in order of their retention times. We rapidly realized that the critical analysis of such lists is tedious and impractical. Therefore we have preferred a classification of the identified constituents into six groups according to their chemical functionality: hydrocarbons, alcohols, aldehydes and ketones, esters, phenols and aromatic ethers, and a sixth group including cyclic ethers together with miscellaneous compounds. Table I lists the rose varieties in which one component exceeds 19-20%. Table II indicates, for the 27 cultivars, the composition with regard to the 4 main functional groups

To symplify the presentation of the results we have chosen a graphical illustration of a flower with six petals whose individual lengths are proportional to the functionalities of the volatile constituents (Figure 1). The quantities are expressed in % of the total amount of volatiles detected in the headspace. On the basis of our systematic observation of these *"Exhalograms or Exograms"* we would like to propose a classification of odoriferous roses consisting of several types:

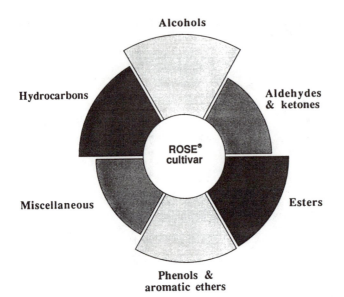

Figure 1. "Exogram" showing the distribution of rose headspace volatile constituents according to their chemical functionalities.

Figure 2. Rose varieties of "hydrocarbon type".

Hydrocarbon Types

The major functional group of these varieties is represented by terpenic, homoterpenic and sesquiterpenic hydrocarbons. Among the varieties examined we found three cultivars whose volatile headspace contains hydrocarbons to the extent of more than 40% (Figure 2):

MONICA (Tantau 1986). This rose presents very elegant buds with a copper-orange yellow colour. The major volatile components are hydrocarbons: *(E)*-4,8-dimethyl-1,3,7-nonatriene (22.3%), caryophyllene (21.5%), various sesquiterpenes (12.5%), and aromatic ethers (13.8%) (orcinol dimethylether; 10.5%). Some esters are also present: phenylethyl acetate (3.1%), *(Z)*-3-hexenyl acetate (1.5%) and hexyl acetate (1.2%). Phenylethanol is only present in small quantities (1.3%). The olfactive contribution of β-ionone (2.0%) is evident.

SONIA MEILLAND (syn. SWEET PROMISE, var. Meihelvet, Meilland 1973). This hybrid tea rose has China pink petals with salmon highlights and is distinguished by its pronounced fruity smell (Origin: "ZAMBRA" x "SEEDLING"). The hydrocarbon group (46.1%) mainly consists of germacrene-D (43.1%) whereas phenylethanol (20.9%) and phenylethyl acetate (5.7%) are the principal alcohol and ester. Miscellaneous minor compounds such as hexanal (0.5%), (Z)-2-hexenal (0.5%), decanal (0.5%) and anisole (0.6%) are important contributors to the overal odor.

CHARLES DE GAULLE (var. Meilanein, Meilland 1975). The volatiles of this mauve-blue rose mainly consist of a multitude of sesquiterpenic hydrocarbons (40.7%) including germacrene-D (8.9%), β-caryopyllene (2.1%) and γ-muurolene (2.7%). The typical smell can be explained by the presence of hexyl acetate (11.7%), (Z)-3-hexenyl acetate (9.4%), a small quantity of citronellol (3.4%) but more probably by trace constituents such as anisole (0.9%) and rose oxides (1.5%).

Alcohol Types without Orcinol Dimethylether.

The classical "old" varieties of roses contain between 35% and 85% of essential alcohols such as phenylethanol, citronellol, nerol and geraniol, but the cumulative quantity of these alcohols and various esters remain between 75 and 85%. These varieties contain only small quantities of orcinol dimethyl ether. The following five varieties belong to this class (Figure 3):

Rosa rugosa rubra. This variety characterized by large, simple red bordeaux petals has a headspace containing 81% of alcohols (phenylethanol: 53.9%, citronellol: 16.8%, geraniol: 5.6%). Only 2.1% of phenylethyl acetate is present. Among the typical trace constituents are benzaldehyde (0.6%), geranial (0.5%), nonanal (0.3%), decanal (0.2%), rose oxides (0.4%) and methyleugenol (0.6%).
     The "classical" rose smell of this variety is easily explained by the high "basic" alcohol content, with the honey, mild and warm notes of phenylethanol predominating.
Rosa muscosa purpurea. This variety is not very different from *R. rugosa rubra.* It contains 70.5% of alcohols (phenylethanol: 29.9%, geraniol: 20.6%, citronellol: 16.3%). The phenylethyl acetate content is higher (7.8%). The presence of myrcene (6.6%) accounts for a more balsamic, resinous smell.

Rose à parfum de l'Hay (Gravereaux 1901). This rose is characterized by red carmine petals and the presence of 56.5% of alcohols (phenylethanol: 23.2%, citronellol: 18.9%, geraniol: 12.5%). Richer in esters (23.1%) than *R. mucosa purpurea* it contains (Z)-3-hexenyl (16.6%), citronellyl and neryl acetates which explains the green, fruity and geranium notes of its perfume. Small quantites (2.2%) of hydrocarbons (α-cubebene and (E)-ocimene) are also present. Geranial (1%) contributes to the lemon smell. Isoamyl acetate (0.2%) and rose oxides (0.1%) are also detected.

Rosa gallica. This classical rose, also known as Rose de Provins (var. officinalis) since the 16th century, has bright red petals. Containing 47.5% of alcohols (phenylethanol, 41.8%, nerol, 2.6%) it is relatively rich in esters (28.1%) (geranyl acetate: 18.1%, phenylethyl acetate: 7.7%) and contains a non-negligible quantity of sesquiterpenes (germacrene: 4.4%, bourbonene: 2.1%).

Rosa damascena. Well known by the perfumers as the origin of Bulgarian rose oils, this classical variety can be included in the first group despite its higher content in esters than alcohols. In line with the other varieties of this group, the volatiles lack the presence of orcinol dimethylether. Alcohols represent 37.9% of the volatiles (phenylethanol: 14.1%, nerol: 11.9%, geraniol: 9.7%) and esters 45.5% ((Z)-3-hexenyl acetate: 29.7%, phenylethyl acetate: 5.8%, hexyl acetate: 5.1%). The good balance between characteristic alcohols and esters (total: 83.4%) accounts for the quality of its essence. Geranial is present at a significant level (1.1%).

Alcohol Types containing Orcinol Dimethylether (Figure 4)

MARGARET MERRIL (Harkness 1978). This floribunda waxy white flower with a nuance of satin-like pink color is considered as the sweetest and most deliciously perfumed white flower (Origin: "SEEDLING" x "PASCALI"). Very rich in alcohols (62.5%) its headspace surprisingly contains hexanol as the main constituent (39.5%), followed by citronellol (16.3%), phenylethanol (3.7%) and geraniol (2.3%). The principal ester is hexyl acetate (5.2%). Hydrocarbons are: β-caryophyllene (4.2%), β-pinene (1.6%) and limonene (0.5%). A small quantity of rose oxides (0.9%) is also present.

WESTERLAND (W.Kordes 1969). Flower with pink petals, yellow at the base. The major group of volatiles is made up of alcohols (45.7%): nerol (17.2%), phenylethanol (14.3%) and geraniol (14.2%). Among the esters (29.3%) the presence of phenylethyl acetate (16.0%), hexyl acetate (4.6%), geranyl acetate (4.3%) and neryl acetate (2.7%) should be mentioned. Other minor but olfactively important constituents are: methyleugenol (2.9%), benzaldehyde (0.05%) and β-ionone (0.05%).

CHATELAINE DE LULLIER (var. Meipobil, Meilland 1987). This large pink flower with silvery tones is nicely balanced with respect to alcohols (37.6%), esters (23.0%) and aromatic ethers (23.1%); the main constituents being: phenylethanol (26.1%), citronellol (11.5%), phenylethyl acetate (9.0%), (Z)-3-hexenyl acetate (6.8%), geranyl acetate (2.3%) and orcinol dimethylether (23.0%).

Ester Types

"Ester" varieties are those whose total ester headspace content is higher than 30%. Nearly all of them contain moderate quantities of alcohols (<25%) and orcinol dimethylether (<30%). (Figure 5).

BARONNE E. DE ROTHSCHILD (var. Meigriso, Meilland 1969). This variety (tea hybrid) has distinct double flowers carrying purple-Solferino petals with a silvery reflection on the underside. This rose can be considered the champion variety with respect to ester content (91.1%): (Z)-3-hexenyl acetate (45.1%) and hexyl acetate (32.9%).

SUTTERS GOLD (Swim 1950). This hybrid tea variety has copper yellow petals veined by orange and carmine colors (Origin: "CHARLOTTE ARMSTRONG" x "SIGNORA"). Its creation commemorated the centenary of the discovery of the gold mines in Sutter Creek, California. It is characterized by an oriental rose smell, slightly fruity with cherry and cognac undertones. Esters are predominant (75.5%) with an homogeneous distribution: (Z)-3-hexenyl acetate (23.7%), hexyl acetate (26.8%), phenylethyl acetate (16.3%) and geranyl acetate

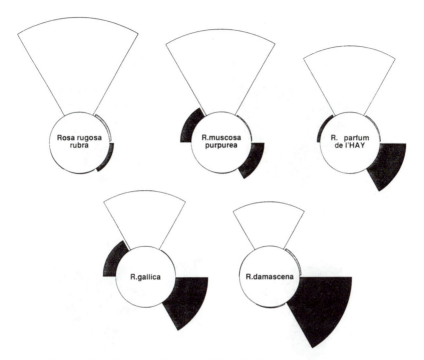

Figure 3. Rose varieties of "alcohol type", non-containing aromatic ethers.

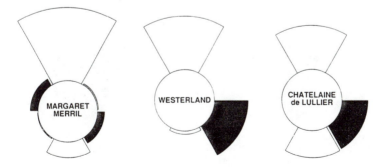

Figure 4. Rose varieties of "alcohol type" containing orcinol dimethylether.

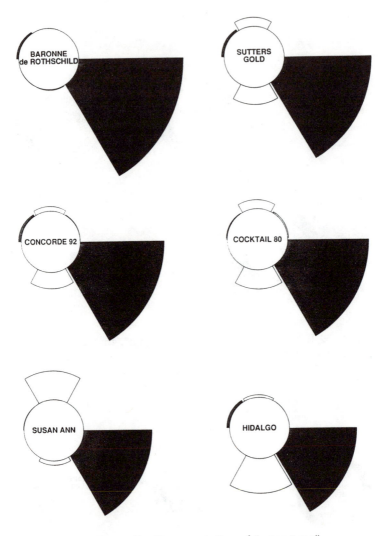

Figure 5a. Rose varieties of "ester type".

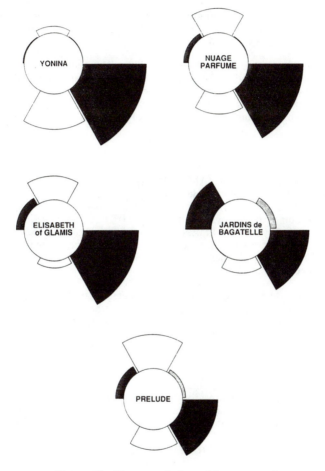

Figure 5b. Rose varieties of "ester type".

(4.2%). Orcinol dimethylether is distinctly present (14.2%) and non-negligible proportions of alcohols have been identified: phenylethanol (3.2%), nerol (2.7%) and citronellol (1.8%). Traces of rose oxides have also been detected (0.18%).

CONCORDE 92 (var. Meidorsun, Meilland 1992). Petals of the flower are coloured dark lemon yellow. Its odor is characterized as green, floral and earthy-woody. Esters represent 73.2% of total volatiles: (Z)-3-hexenyl acetate (48.8%), phenylethyl acetate (16.7%) and hexyl acetate (7.4%). A moderate proportion of orcinol dimethylether (13.4%) and smaller proportions of phenylethanol (2.6%) and germacrene-D (1.8%) are present.

COCKTAIL 80 (var. Meitakilor, Meilland 1980). Warm and bright yellow petals with an odor characterized as floral, rosy, fruity and green. Esters constitute the main functional group (70.9%): phenylethyl acetate (59.8%), geranyl acetate (6.3%) and (Z)-3-hexenyl acetate (3.4%). The proportion of orcinol dimethylether is 16.1%, with a smaller quantity of phenylethanol (5.9%) and detectable traces of β-ionone (0.3%).

SUSAN ANN (syn.: SOUTHAMPTON, Harkness 1972). This floribunda flower has an orange, apricot shade color. (Origin: "SEEDLING" x "YELLOW CUSHION"). It is characterized by its opulence in esters (59.7%) and alcohols (24.4%), particularly by phenylethyl acetate (42.0%) and phenylethanol (22.1%). Other esters are hexyl acetate (9.0%) and (Z)-3-hexenyl acetate (6.9%). Orcinol dimethylether is a minor constituent (3.3%). Trace compounds which can influence the smell of this variety are methyleugenol (0.1%), benzaldehyde (0.1%), nonanal (0.3%) and decanal (0.4%).

HIDALGO (var. Meitulandi, Meilland 1980). Petals of this turbinate flower have a red velvet color. Esters (58.1%) and aromatic ethers (29.7%) are the main constituents: hexyl acetate (43.2%), (Z)-3-hexenyl acetate (14.5%) are accompanied by benzyl methylether (13.8%) and orcinol dimethylether (9.9%). Phenylethanol (1.7%) and nerol (0.6%) are accompanied by rose oxides (0.9%) and various sesquiterpenes (2.5%).

YONINA (var. Meispola, Meilland). The petals are white, lightly shaded with pink. Esters are again the main group (55.8%): (Z)-3-hexenyl acetate (26.5%), hexyl acetate (26.2%) and geranyl acetate (2.1%). Orcinol dimethylether is again a major componant (29.0%) while the level of citronellol is significant (5.3%).

NUAGE PARFUME (syn.: DUFTWOLKE, FRAGRANT CLOUD, Tantau 1963). This hybrid tea variety with a dark fire-red flower is classified as *World's Favourite Rose* N°3. (Origin: "SEEDLING" x "PRIMA BALLERINA"). Proportions of esters and alcohols are nearly the same as in SUSAN ANN: 48.8% and 21.1% respectively. The most abundant esters are phenylethyl acetate (19.3%), hexyl acetate (12.3%), (Z)-3-hexenyl acetate (9.7%) and citronellyl acetate (7.1%). Alcohols are phenylethanol (14.8%) and citronellol (5.8%). This variety contains non-negligible quantities of olfactively interesting constituents such as p-vinylanisole (11.0%), β-ionone (0.7%), rose oxides (0.2%) and estragol (0.1%).

ELISABETH OF GLAMIS (syn.: IRISH BEAUTY, McGredy 1964). This floribunda variety has large, double flowers with pink, carmine salmon petals with a gold tone. Esters are the main constituents (48.6%): phenylethyl acetate (24.5%), hexyl acetate (20.4%) and *(Z)*-3-hexenyl acetate (2.9%). Phenylethanol represents 19.5% and *p*-methoxytoluene 4.4%.

JARDINS DE BAGATELLE (var. Meimafris, Meilland 1986). This white rose with pink undertones is famed for its typical rosy, faintly lemon-like smell. Esters constitute the main group (34.2%), nearly equally distributed between two constituents: hexyl acetate (15.5%) and *(Z)*-3-hexenyl acetate (14.0%). Among hydrocarbons figure β-caryophyllene (16.2%). Orcinol dimethylether represents 9.6% of total volatiles. This rose is the champion of β-ionone-containing roses with a level of 5.7%. Rose oxides are also present (0.5%).

PRELUDE (var. Keimove, Meilland). This mauve rose has a powerful odor, having a good balance between alcohols (29.2%) and esters (30.9%): geraniol (14.3%) and nerol (12.7%) on one hand, and four esters on the other hand: *(Z)*-3-hexenyl acetate (17.5%), geranyl acetate (5.6%), hexyl acetate (4.1%) and phenylethyl acetate (2.4%). Other major components orcinol dimethylether (17.0%), germacrene-D (5.5%) and geranial (2.3%).

Aromatic Ether Types

These varieties contain more than 40% of orcinol dimethylether (Figure 6):

SYLVIA (W.Kordes, 1978). Characterized by salmon-colored pink petals, this rose is very rich in orcinol dimethylether (67.3%), it also contains anisole (0.1%) and *p*-vinylanisole (1.1%). Quantitatively hydrocarbons are the second most important constituents (germacrene-D: 6.9%, cubebene: 2.9%). Esters are almost absent but typical notes are certainly explained by the presence of a relatively high proportion of β-ionone (2.7%) and the presence of rose oxides (0.54%), nonanal (0.9%) and decanal (0.9%).

YOUKI SAN (var. Meidonq, Meilland 1965). This tea hybrid pure white rose is also very rich in orcinol dimethylether (55.8%), anisole (1.4%) and *p*-vinylanisole (0.6%). It contains small amounts of esters (hexyl acetate: 3.3%, *(Z)*-3-hexenyl acetate: 1.6%) and almost no alcohols. The typical smell of this rose could be due to the presence of nearly 1% of phenylacetaldehyde.

GINA LOLLOBRIGIDA (var. Meilivar, Meilland 1989). This ardent yellow-gold rose is also characterized by a high level of orcinol dimethylether (52.5%) and the presence of anisole (1%). Quantitatively esters are the second most important constituents (18.6%): *(Z)*-3-hexenyl acetate (11.4%), geranyl acetate (1.6%), citronellyl acetate (1.1%) and neryl acetate (0.5%). "Classical" alcohols such as nerol (9.2%) and phenylethyl alcohol (2.3%) balance the perfume of the flower.

PAPA MEILLAND (var. Meicesar, Meilland 1963). This dark velvet tea hybrid red rose has a smell which has been characterized as "very classical rose" with a lemon-peel note. Its high level of aromatic ethers (44.8%) is well balanced by the moderate amounts of alcohols (31.5%) and esters (16.2%). The major constituent is still orcinol dimethylether (40.3%) and is accompanied by benzyl methylether (2.7%) and *p*-vinylanisole (1.6%). Citronellol (29.4%) and phenylethanol (1.8%), hexyl acetate (12.6%) and citronellyl acetate (3.1%) constitute the main components with a non-negligible proportion of rose oxides (1.5%) and β-caryophyllene (1.0%).

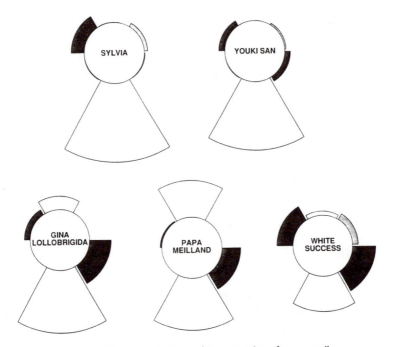

Figure 6. Rose varieties of "aromatic ether type".

WHITE SUCCESS (var. Jelpirofor, Meilland). Also marketed in Japan as MADAME SACHI SUSUKI, this rose has petals delicately shading from white to pale-cream yellow. Surprisingly its odor is characterized as "jasmine-like". With a better balance between the various functional groups than Papa Meilland, it contains in decreasing order of importance orcinol dimethylether (29.6%), esters (20.0%): phenylethyl acetate (10.1%) and (Z)-3-hexenyl acetate (7.2%), sesquiterpenes (14.1%): germacrene (7.89%) and β-caryophyllene (3.0%), acetone (5.3%) and phenylethanol (2.3%).

Although some samplings have been duplicated using the same varieties or even on the same flowers at two days intervals, during their optimum blooming, at present we can neither guarantee the strict reproducibility of these quantitative results, nor indicate their possible deviations. The emitted volatiles evidently depend on the maturity of the flower and on a multitude of parameters, e.g. outdoor temperature and humidity, solar radiation or the orientation of the stems. However, we have observed that results published independently by other laboratories (8) (21), on identical botanical species did not significantly differ from ours. Moreover, we propose this type of quantitative presentation as an efficient tool for simple characterization of flower headspace. The number of "exogram petals" can evidently be adapted to different and perhaps more significant functional groups. Indeed,by dividing the concentration of the individual constituents by their respective olfactive thresholds one should obtain a far more significant "odorgram", based on the real odor value (17) (22). This type of graphical analysis should help to highlight the contribution of minor constituents which are essential for the creation of new, original and more "natural" perfumes.

Acknowledgments. The authors thank Dr Serge Gudin (Selection MEILLAND, Antibes), Dr L.Peyron (Grasse), Prof. R. Spichiger (Jardin Botanique de Genève), Mr. J.M. Mascherpaz (Centre horticole de Lullier, Genève) and Mr. E. Tschanz (Roseraies TSCHANZ, Lausanne) who provided them with rose varieties or allowed to sample headspace volatiles.

Literature Cited.

1. Kaiser, R.; Lamparsky, D. *Parfums Cosmet. Aromes*, **1977**, *17*, 71-79.
2. Tsuneya, T.; Ikeda, N.; Shiga, M.; Ichikawa, N. *VIIth Int. Cong. Essential Oils, Kyoto (Japan)*, **1977**, *Oct.7-11*, 454-457.
3. Mookherjee, B.D.; Trenkle, R.W.; Wilson, R.A. *Pure Appl. Chem.* **1990**, *62*, 1357-1364.
4. Bicchi, C.; d'Amato, A.; David, F.; Sandra, P. *Flavour Fragrance J.* **1987**, *2*, 49-54.
5. Bicchi, C.; Joulain, D. *Flavour Fragrance J.* **1990**, *5*, 131-145.
6. Joulain, D. *Flavour Fragrance J.* **1987**, *2*, 149-155.
7. Surburg, H; Güntert, M. *Haarmann & Reimer Contact*, **1991**, *N°51*, 12-17.
8. Brunke, E.J.; Hammerschmidt, F.J.; Schmaus, G. *Dragoco Rep. (Germ.Edit.)*, **1992**, *N°1*, 3-31.
9. Flament, I.; Sauvegrain, P.; Chevalier, C. *Huitièmes Journées Internationales Huiles Essentielles, Digne-les-Bains (France)*, 31.8-2.9, **1989**.
10. Kovats, E. *J.Chromatog.* **1987**, *406*, 185-215.
11. Naves, Y.R. *Helv.Chim.Acta*, **1949**, *32*, 967-968.
12. Seidel, C.F.; Stoll, M. *Helv.Chim.Acta*, **1959**, *42*, 1830-1844.
13. Seidel, C.F.; Felix, D.; Eschenmoser, A. *Helv.Chim.Acta*, **1961**, *44*, 598-606.
14. Büchi, G.; Kovats, E.; Enggist, P.; Uhde, G. *J.Org.Chem.* **1968**, *33*, 1227-1229.
15. Ohloff, G.; Giersch, W.; Schulte-Elte, K.H.; Kovats, E. *Helv.Chim.Acta*, **1969**, *52*, 1531-1536.
16. Demole, E.; Enggist, P.; Säuberli, U.; Stoll, M.; Kovats, E. *Helv.Chim.Acta*, **1970**, *53*, 541-556.
17. Ohloff, G. *Perfum.Flavor.* **1978**, *3*, 11.
18. Flament, I. *Cosmetics & Toiletries Manufacture*, **1991-92**, 114-122.
19. Kaiser, R. In *Perfumes, Art Science & Technology, Editors Müller, P.M.; Lamparsky, D. Publisher: Elsevier Applied Science, London and New-York*, **1991**, 213-250.
20. Mookherjee, B.D.; Trenkle, R.W.; Wilson, R.A.; Zampino, M.; Sands, K.P.; Mussinan, C.J. In *Flavors and Fragrances: A World Perspective. Proc. 10th Int.Cong.Essent.Oils, Fragrances and Flavors, Washington, DC, U.S.A.*, **1986**, Nov.16-20, 415-424.
21. Eugster, V.C.H.; Marki-Fischer, E. *Angew. Chem. Int. Ed. Engl.* **1991**, *30*, 654-672.
22. Rothe, M. *Die Nahrung*, **1976**, *20*, 259-270.

RECEIVED August 19, 1992

Chapter 20

Flower Scent of Some Traditional Medicinal Plants

E.-J. Brunke, F.-J. Hammerschmidt, and G. Schmaus

Dragoco Research Laboratories, D–W–3450 Holzminden, Germany

The volatile constituents of the inflorescences of the following medicinal plants relevant to the perfume and flavor industry were analysed: Chamomile (*Matricaria recutita* L.), Roman chamomile (*Chamaemelum nobile* (L.) All.), Lavender-Cotton (*Santolina chamaecyparissus* L.), Valerian (*Valeriana officinalis* L.), and Meadowsweet (*Filipendula ulmaria* L. Maxim). Comparative analyses of headspace concentrates of unpicked flowers and of corresponding hydrodistillates showed significant differences in most cases.

Medicinal plants have played an important role throughout the history of Man. Most of these plants bear flowers, admired for their form, scent and medicinal properties. These flowers represent links between medicine, aroma therapy, and aesthetics.

Chamomile (*Matricaria recutita* L.)

Chamomile (*Matricaria recutita* L.; syn. *Chamomilla recutita* (L.) Rauschert), family *Asteraceae (Compositae)*, is an annual herb with white, yellow centered flowers, native to Southern and Eastern Europe as well as Asia Minor. Because of its cultivation as medicinal plant, it has become a real cosmopolitan. The world production of dried chamomile flowers, the most important part of the plant, represents more than 6500 tons per year (*1*).

The medicinal and cosmetic properties of chamomile, known in Europe since antiquity and acknowledged in modern times, depend on extractives but also on the volatiles of chamomile flowers. The essential oil of chamomile is an expensive ingredient for pharmaceutical and cosmetic preparations; it is used also in the fragrance and flavor industry.

0097–6156/93/0525–0282$06.00/0

An olfactory comparison between fresh and dried chamomile flowers shows significant differences. Only the living chamomile flower has the typical fresh-floral, herbaceous and fruity notes.

Research reports cover detailed results on constituents of the essential oil and extracts of chamomile flowers as well as their therapeutical and cosmetic properties (*1, 2*). Nothing, however, has been published on the volatile constituents emitted by the living chamomile plant.

Because of its attractive and characteristic odor, we have analysed the headspace of chamomile growing wild in the surroundings of Holzminden. For the determination of the chamomile chemotype, we have also performed steam distillation with the same plant material. The GC-MS analysis of the essential oil of chamomile obtained hereby showed a high bisabolol oxide A content, which is typical for the bisabolol oxide A chemotype, the more common type in central Europe (*1*). Other constituents with a bisabolane structure were bisabolol oxide B, bisabolone oxide A, and α-bisabolol. Chamazulene with its dark blue color is formed as an artifact during the steam distillation (Table I).

For the analysis of unpicked flowers, we generally use the Closed Loop Stripping method. In an application of this technique, unpicked flowers are put into a glass flask. Cotton-wool is placed around the stem to provide a virtually closed system. Air is circulated for about 2 - 4 hours by a suction/pressure pump (Flow: 100-150 ml/min) to transport the volatiles emitted by the flowers to an adsorption tube filled with 20 - 50 mg of active charcoal (Klimes Company, CH-8600 Dübendorf). The volatiles released by the flower are retained on the adsorbent, while the air circulates through pump, cleaning filter (active charcoal) and glass vessel containing the plant material. After trapping, the volatiles are immediately desorbed with carbon disulfide or diethyl ether and subsequently analysed by GC and GC/MS. [Conditions: 1. GC: a) 60 m DB-WAX capillary column; Temp. Progr.: 50 - 200 °C, 4°C/min; b) 30 m DB-1 column; Temp. Progr.: 50 - 240 °C, 4°C/min; 2. GC/MS: 60 m DB-WAX and 30 m DB-1 column; Temp. Progr. 50 - 240 °C, 4°C/min; Ionisation energy: 70 eV].

The comparison of steam distillate and headspace concentrate of chamomile flowers shows dramatic differences. The main components of the essential oil are only found as mere trace substances in the headspace, or they are not even detectable. On the other hand, the headspace concentrates of unpicked chamomile flowers contain a number of volatile monoterpenoids such as 1,8-cineole, p-cymene and artemisia ketone or sesquiterpene hydrocarbons such as caryophyllene, E-ß-farnesene and ß-selinene at higher concentrations. A number of short chain aliphatic esters like ethyl- and propyl-2-methyl butyrate, methyl tiglate, ethyl- and butyl isovalerate and cis-3-hexenyl acetate also play an important role in the natural smell of chamomile, as could be shown by perfumistic reconstitutions of the chamomile flower scent.

Artifact Formation During Headspace Sampling.

In the chamomile headspace we found 2,6-dimethyl-1,3,5,7-octatetraene as well as 2,6-dimethyl-3,5,7-octatrien-2-ol as a mixture of Z/E isomers. These substances have been found previously in 1977 by Kaiser and Lamparsky in the headspace of

Table I: **Volatile constituents of Chamomile flowers**
            (*Matricaria recutita* L., syn.: *Chamomilla recutita* (L.) Rauschert)
            Comparison of the hydrodistillate with the "Closed Loop Stripping"
            sample obtained from living flowers

Peak No.	Compound	GC - Area - %	
		Hydrodistillate fresh flowers	Closed Loop Stripping living flowers
3	α-Pinene	trace	1,26
4	Ethyl butyrate, 2-methyl-	0,17	6,15
5	Ethyl isovalerate	-	0,12
7	ß-Pinene	-	0,14
8	Sabinene	0,22	0,39
9	Propyl butyrate, 2-methyl-	trace	2,22
10	Myrcene	trace	0,75
11	Methyl tiglate	-	0,12
12	Limonene	trace	0,97
13	1,8-Cineole	0,34	8,75
14	Butyl isovalerate	-	0,18
15	cis-ß-Ocimene	0,08	-
16	trans-ß-Ocimene	0,70	-
17	Hexyl acetate	-	0,23
18	p-Cymene	0,07	22,23
20	cis-3-Hexenyl acetate	trace	6,95
21	Hexyl propionate	-	0,17
22	5-Hetenone-(2), 6-methyl-	trace	0,24
23	Artemisia ketone	0,51	10,34
25	cis-3-Hexenol	-	0,31
26	cis-3-Hexenyl propionate	-	0,37
27	trans-2-Hexenol	0,10	0,45
30	1,3(E),5(Z),7-Octatetraene, 2,6-dimethyl-,	-	0,12
31	1,3(E),5(E),7-Octatetraene, 2,6-dimethyl-,	-	0,12
32	Artemisia alcohol	0,39	0,51
34	Benzaldehyde	-	trace
37	Lavandulyl acetate	-	0,32
38	ß-Elemene	-	0,24
39	Terpinenol-(4)	trace	

Table I Cont.

Peak No.	Compound	GC - Area - %	
		Hydrodistillate fresh flowers	Closed Loop Stripping living flowers
40	ß-Caryophyllene	0,09	5,10
41	Methyl benzoate	-	0,14
43	E-ß-Farnesene	2,92	4,90
44	Lavandulol	-	trace
45	α-Humulene	-	0,33
46	α-Terpineol	-	0,28
48	Germacrene-D	1,48	-
49	ß-Selinene	trace	15,97
50	Bicyclogermacrene	0,64	0,92
51	3(E),5(Z),7-Octatriene-2-ol, 2,6-dimethyl-,	-	0,24
52	3(E),5(E),7-Octatrien-2-ol, 2,6-dimethyl-,	-	0,37
54	Bisabololoxide B	4,35	0,19
55	Bisablolonoxide A	4,10	1,15
56	α-Bisabolol	0,51	-
58	Chamazulene	23,35	-
59	Bisabololoxide A	57,68	0,17

hyacinth flowers (3). The substances have been described as unstable so that they are only detectable in headspace samples but not in extracts or distillates of the same plant material. These alcohols and hydrocarbons have been found in the headspace of numerous flowers, sometimes in different concentrations depending on diurnal rhythms (4). We have also detected these substances in several other fragrant flowers during our headspace work, but now we have to revise these results.

Recently, we compared adsorption materials such as charcoal and Tenax from different sources and could find the substances mentioned only by using one type of charcoal adsorption traps. With use of another charcoal quality as well as Tenax qualities, these substances could be detected only at very low or zero concentrations. We also observed that the dimethyloctatetraenes as well as the corresponding trienols occur in combination with either lower or zero concentrations of Z- and E-ß-ocimene. To examine whether these substances could be artifacts, we performed Closed Loop Stripping experiments with a Z/E-ß-ocimene test mixture in high dilution, using Klimes traps. After desorption it could be demonstrated by GC/MS that the dimethyloctatetraenes and -octatrienols are produced on the active charcoal surface. In addition, we detected 1,5-p-menthadienol-(8) and 1(7),2-p-menthadienol-(8), formal cyclization products of the octatrienols. In a second model experiment, we performed Closed Loop Stripping with the Z/E-octatrienols and found formation of the p-menthadienols at the active charcoal surface during headspace sampling.

These results clearly show that the possibility of artifact formation should always be taken into account when using standard headspace methods.

Roman Chamomile (*Chamaemelum nobile* (L.) All.)

Roman chamomile (*Chamaemelum nobile* (L.) All., syn.: *Anthemis nobilis* L.) is a perennial herb naturally occurring in the Mediterranean countries and of traditional medicinal use. The essential oil with its dry fruity smell is of importance to the fragrance and flavor industry. Main constituents are different esters of (Z)-2-methyl-2-butenoic acid (angelic acid) and other short chain aliphatic esters (5-13). The relative concentrations of the constituents differ widely between samples of different origin, i.e. the isobutyl angelate content varies between 4 and 40 %. Terpenoids only occur in very low concentrations. Recently described as minor constituents were the esters of 3-hydroxy-2-methylidene butyric acid (8, 9) as well as some novel diesters, ketones and p-menthane derivatives (10-12) (Table II).

In contrast to the detailed knowledge on the essential oil of commercial origin, only one publication concentrates on a static headspace analysis of picked and chopped Roman chamomile flowers (13). Isobutyl angelate was found to be the main component. The authors described 13 constituents, among others isoamyl and hexyl butyrate. In our work the presence of these substances could not be verified.

We performed headspace sampling with unpicked Roman chamomile flowers from cultivation in Holzminden. As could be shown by comparative GC/MS investigations, the compositions of the headspace and of steam distilled samples are similar. Different concentration ratios of the short chain aliphatic esters, lower concentrations of some oxygenated monoterpenes and the absence of the 3-hydroxy-2-methylidene butyric acid esters in the headspace are the only differences.

Lavender Cotton (*Santolina chamaecyparissus* L.)

Santolina chamaecyparissus L. is native to Mediterranean countries. Because of its beautiful shape and yellow flowers as well as its strong aromatic smell and therapeutical properties, *Santolina chamaecyparissus* has been cultivated as a decorative and medicinal plant for many centuries. Recently, production and commercialisation of its essential oil have been started in southern France. The smell of this oil is composed of herbaceous, fresh camphoraceous and spicy aspects, reminiscent of vermouth and Roman chamomile. Because of its interesting smell and the incomplete knowledge of its constituents, we have analysed the essential oil of *Santolina chamaecyparissus* in more detail (Table III).

GC-analysis shows the complex composition of the oil. One of the main components is artemisia ketone (10-40 %); additional irregular monoterpenes are artemisia alcohol, yomogi alcohol and santolinatriene (*14-20*). Isoartemisia ketone, described as a constituent (*16*), could not be confirmed. We have found chrysanthemol as a new constituent. This interesting substance is being discussed as a biogenetic precursor for irregular terpenes of the artemisyl, lavandulyl and santolinyl family (*21*). We have also found a number of oxygenated regular monoterpenes as well as some aliphatic and aromatic components, all new for santolina oil (*20*).

The fraction of sesquiterpenoids contained a relatively large amount of functionalized sesquiterpenes which were unidentified up to now. We isolated the main component of this fraction (10-18 % of the oil) by distillation and preparative gas chromatography. NMR experiments, mainly homo- and heteronuclear correlations, demonstrated that the substance is longiverbenon. This rarely occurring natural substance has been described first by Uchio in 1977 as a constituent of *Tanacetum vulgare* L. (*22-24*). Other tricyclic sesquiterpenoids which could be found for the first time in santolina oil are vulgarone A, cis- and trans-longiverbenol and 4-oxo-α-ylangene (*20*). All of these substances are also uncommon natural compounds.

For the headspace analysis of *Santolina chamaecyparissus* we have used unpicked flowers and leaves of santolina plants cultivated in Holzminden in a suitable micro-climate. Both parts of the plant are odorous and their headspaces contain the same substances as the steam distillate but in different ratios. Artemisia ketone plays an important sensory role in the headspace: The vapor phase above the leaves contains between 30-40 % and that of the flowers more than 60 % of artemisia ketone (its content in the corresponding essential oil is 10 %). Further constituents contributing to the typical smell of santolina leaves and flowers seem to be 1,8-cineole, cis-3-hexenol and cis-3-hexenyl acetate. Due to their low vapor pressure, the functionalized sesquiterpenoids proved to be only minor constituents in the headspace samples.

Table II: **Volatile constituents of Roman Chamomile flowers**
(*Anthemis nobilis* L.: syn.: *Chamaemelum nobile* (L.) All.)
Comparison of the hydrodistillate with the "Closed Loop Stripping" sample obtained from living flowers

Peak No.	Compound	GC Area %	
		Hydrodistillate fresh flowers	"Closed Loop Stripping" living flowers
5	Ethyl isobutyrate	trace	-
6	Isobutyl acetate	0,02	0,51
8	α-Pinene	1,05	1,18
10	Camphene	0,39	1,54
11	Isobutanol	-	0,31
12	Isobutyl isobutyrate	0,75	2,77
14	Isoamyl acetate	0,93	1,82
15	2-Methylbutyl acetate		
18	Isobutyl butyrate	0,04	trace
20	Ethyl angelate	0,08	trace
21	Isobutyl methacrylate	0,33	1,74
22	Isobutyl 2-methylbutyrate	-	0,45
23	2-Methyl-2-propenyl-isobutyrate	0,88	-
24	Isoamyl propionate	0,06	-
25	Isoamyl isobutyrate	3,13	0,50
26	2-Methylbutyl isobutyrate		2,55
28	1,8-Cineole	0,12	-
30	3-Methylpentyl acetate	0,94	2,17
32	2-Methyl-2-butenyl acetate	-	trace
33	Propyl angelate	1,04	1,58
34	2-Methyl-2-propenyl-methacrylate	0,81	0,28
36	2-Methyl-2-butenyl isobutyrate	0,15	-
37	Isoamyl methacrylate	1,53	1,24
38	2-Methylbutyl 2-methyl-butyrate	-	trace
39	Isobutyl angelate	4,03	28,52
40	3-Methylpentyl propionate	0,10	-
42	3-Methylpentyl isobutyrate	12,54	2,14
43	cis-3-Hexenyl acetate	-	0,38
44	3-Methylpentanol	0,19	0,35
46	n-Butyl angelate	0,37	0,84
48	2-Methyl-2-propenyl angelate	13,14	7,64
50	3-Methylpentyl methacrylate		0,64
52	Isoamyl angelate	17,54	7,33
53	2-Methylbutyl angelate		15,01
54	3-Methylpentyl isovalerate	0,04	trace
58	n-Amyl angelate	0,07	trace
59	1-Octen-3-ol	0.05	-

Table II Cont.

Peak No.	Compound	GC Area %	
		Hydrodistillate fresh flowers	"Closed Loop Stripping" living flowers
62	2-Methyl-2-propenyl tiglate	0,80	0,35
64	2-Methylbutyl tiglate	0,17	-
66	3-Methylpentyl angelate	22,69	10,61
69	Benzaldehyde	0,02	-
71	Camphor	0,02	-
76	Isobutyric acid	0,03	-
81	3-Methylpentyl tiglate		-
82	Pinocarvon	2,37	0,79
87	Myrtenal	0,67	-
89	β-Farnesene	0,10	-
90	Pinocarveol, trans-	4,36	0,71
94	α-Terpineol	0,03	-
96	Borneol	0,20	-
102	Myrtenyl isobutyrate	0,05	1,04
103	ar-Curcumene	trace	-
104	Isobutyl 2-hydroxy-2-methylenebutyrate	0,08	-
105	Benzyl isobutyrate	0,10	-
107	Myrtenol	0,38	trace
109	Myrtenyl 2-methylbutyrate	-	0,45
111	p-Cymenol-(8)	0,05	-
115	Isoamyl 3-hydroxy-2-methylenebutyrate	0,48	-
121	3-Methylpentyl 3-hydroxy-2-methylenebutyrate	0,66	-
129	Nonanoic acid	0,04	-
134	Decanoic acid	0,01	-

Table III: **Volatile constituents of Lavender Cotton** (*Santolina chamaecyparissus* L.)
Comparison of the hydrodistilled herb oil with "Closed loop Stripping"
samples obtained from living leaves and flowers

Peak No.	Compound	GC - Area %		
		Hydrodistillate of leaves	CLS - Samples	
			leaves	flowers
3	α-Pinene	0,30	0,67	0,62
5	Camphene	0,28	3,53	1,36
7	ß-Pinene	2,13	2,92	1,24
8	Sabinene	2,06	3,53	0,88
10	Myrcene	7,38	1,26	0,34
14	Limonene	0,86	0,75	0,34
15	ß-Phellandrene	8,63	6,84	3,34
16	1,8-Cineole	trace	2,10	1,88
20	p-Cymene	0,09	6,40	1,48
22	Terpinolene	0,83	0,09	trace
24	cis-3-Hexenyl acetate	0,04	6,40	1,48
27	Artemisia ketone	8,54	36,37	63,77
30	cis-3-Hexenol	0,02	0,29	0,28
31	Yomogi alcohol	0,35	0,19	0,87
35	4-Isopropenyltoluene	0,01	0,91	0,23
36	cis-3-Hexenyl isovalerate	trace	-	0,33
39	Longipinene	0,48	1,22	2,34
41	Artemisia alcohol	0,49	0,36	1,30
44	Camphor	1,33	2,05	4,29
68	Aromadendrene	0,37	trace	0,73
69	trans-Chrysanthemol	0,48	trace	trace
73	ϒ-Curcumene	2,22	trace	trace
75	Borneol	0,95	0,10	0,98
78	Germacrene D	2,72	trace	0,09
82	Bicyclogermacrene	1,56	trace	0,09
88	ar-Curcumene	0,88	trace	0,05
115	Caryophyllene epoxyde	0,37	-	0,20
145	Longiverbenone	17,36	0,69	1,76

Valerian (*Valeriana officinalis* L.)

Valerian (*Valeriana officinalis* L.) is a perennial herb which is native to Europe and the temperate parts of Asia. The plant grows 1.5 meters high and has white or pink coloured flowers in form of a cyme. The flower presents an interesting strong floral and sweet animal odor. The roots of valerian are used for pharmaceutical preparations. The tincture with its strong and unpleasant smell is popular as a mild sedative. The constituents of valerian roots and its tincture are well known, but nothing has been published on the flower volatiles of *Valeriana officinalis* L.

We performed Closed Loop Stripping and Vacuum headspace experiments with valerian flowers growing wild in the Holzminden area and in addition performed hydrodistillation with the same material. The distillate consisted mainly of isovaleric and valeric acid. Its smell was extremely different from the scent of the living valerian flower (Table IV).

In contrast, the headspace samples from the unpicked valerian flowers contained p-methylanisol and lavandulyl isovalerate as main components. Lavandulyl isovalerate, representing 15-20 % of the different headspace samples is an uncommon natural compound which formerly has only been detected as a trace constituent in *Artemisia fragrans* extracts (*25*) and in the essential oils of several *Lavandula* species (*26, 27*). Further rarely occurring minor components are lavandulyl-2-methylbutyrate and valerate as well as lavandulol. The unpleasant smell of valeric and isovaleric acid could neither be found in the Closed Loop Stripping nor in the vacuum headspace samples.

The comparison of isolation procedures carried out on valerian very impressively proved that steam distillation can deliver completely useless results in the analysis of flower fragrances.

Meadowsweet (*Filipendula ulmaria* L. Maxim.)

Meadowsweet (*Filipendula ulmaria* L. Maxim., syn.: *Spiraea ulmaria* L.) is a perennial herb native to Europe and Northern Asia. It has been introduced to North America as a garden plant. It grows up to 1.3 meters high and has small white to cream-colored flowers in the form of a cyme. The plant prefers wet places, mainly banks or wet meadows. The smell of the flowers dominates in such meadowlands in summertime. It is a sweet-floral scent with vanilla and almond-like aspects. The pharmaceutical usage of meadowsweet flowers was mainly in the field of anti-rheumatic treatments. In former times the flowers were used for flavoring beer and mead. At the end of the last century the German perfume industry tried to obtain commercial products from meadowsweet flowers by steam distillation. The odor quality of the essential oil (spiraea oil) which consists mainly of salicylic aldehyde and methyl salicylate is very different from that of the unpicked flower.

We performed headspace sampling of living meadowsweet flowers as well as steam distillation of the same plant material, growing wild in the surroundings of Holzminden (Table V).

The essential oil obtained hereby consisted mainly of salicylic aldehyde and methyl salicylate, occurring in the plant as glycosides (*28*). Vanillin and heliotro-

Table IV: Volatile constituents of Valerian flowers (*Valeriana officinalis* L.)
Comparison of the hydrostillate and the "Vacuum Headspace" -sample with
"Closed Loop Stripping" samples obtained from living and picked flowers

Peak No.	Compound	GC - Area %			
		Hydro-distillate	"Vacuum-Headspace"	"Closed Loop Stripping"	
				living flowers	picked flowers
4	α-Pinene	-	tr.	tr.	tr.
5	Hexanal	-	tr.	tr.	tr.
7	unknown compound M$^+$ 122	1,99	-	-	-
9	ß-Pinene	-	tr.	tr.	tr.
10	Myrcene	-	tr.	tr.	-
12	Dodecane	-	tr.	1,09	-
14	Limonene	tr.	0,74	1,07	tr.
15	Hexyl acetate	-	1,87	1,59	0,36
16	Isamyl isovalerate	-	1,23	1,34	-
18	cis-3-Hexenyl acetate	tr.	4,51	8,12	0,20
19	trans-2-Hexenyl acetate	-	0,82	1,80	-
21	Hexanol	tr.	1,57	1,45	tr.
23	unknown compound M$^+$ 138	9,70	-	-	-
24	cis-3-Hexanol	-	2,02	0,69	-
25	Nonanal	-	tr.	tr.	-
26	Tetradecane	-	tr.	-	-
28	trans-2-Hexenol	-	1,46	tr.	-
30	p-Methylanisol	-	3,60	20,54	7,51
32	Furfurol	0,50	-	-	-
33	Octyl acetate	-	tr.	1,73	-
35	Pentadecane	-	0,66	2,06	-
36	Benzaldehyde	-	tr.	tr.	-
38	Linalool	-	-	tr.	-
40	Octanol	-	-	1,96	-
42	Methyl benzoate	-	tr.	1,83	0,43
43	E-ß-Farnesene	-	1,22	0,78	tr.
44	Isovaleric acid	36,91	-	-	-
45	Lavandulol	-	1,65	1,24	0,57
46	unknown compound M$^+$ 138	29,33	-	-	-
48	Z,E-α-Farnesene	-	tr.	tr.	tr.
50	E,E-α-Farnesene	tr.	7,65	2,79	2,50
51	Lavandulyl 2-methylbutyrate	-	0,63	2,41	2,32
53	Lavandulyl isovalerate	tr.	15,00	21,16	23,66
55	Valeric acid	13,56	-	-	-
56	Benzene, 1,2-dimethoxy-4-methyl-	-	4,70	4,66	4,49
57	Lavandulyl valerate	-	tr.	tr.	tr.
60	Benzylalcohol	-	2,02	-	-
63	Methyl eugenol	-	2,16	tr.	-

Table V: **Volatile constituents of Meadowsweet flowers** (*Filipendula ulmaria* (L.) Maxim., syn. *Spiraea ulmaria* L.) Hydrodistillate

Peak No.	Compound	GC - Area %
3	Decane	0,05
4	α-Pinene	0,11
6	Hexanal	0,04
8	ß-Pinene	0,02
9	Heptanal	0,15
10	Dodecane	0,02
12	Limonene	0,07
14	trans-ß-Ocimene	0,02
15	Τ-Terpinene	trace
18	Octanal	0,11
19	cis-3-Hexenyl acetate	0,04
22	cis-3-Hexenol	0,12
23	Nonanal	4,75
25	Pentadecane	0,06
26	Decanal	0,07
28	Benzaldehyde	0,13
30	Linalool	2,03
31	Lilac aldehyde (4 isomers)	1,59
32	Undecanal	0,32
36	Lavandulol	0,03
37	Salicylaldehyde	38,20
39	Lilac alcohol (2 isomers)	0,13
41	Methyl salicylate	20,22
42	Tridecanal	0,22
43	Ethyl salicylate	0,19
44	ß-Damascenone	0,76
45	Geraniol	0,53
46	Geranyl acetone	0,16
48	Benzyl alcohol	0,25
50	Nonadecane	0,36
52	ß-Phenylethyl alcohol	0,39
57	Anisaldehyde	0,39
58	Caprylic acid	0,88
59	Hexyl benzoate	0,27
60	Heneicosane	0,36
61	Farnesyl acetone, hexahydro-	0,09
62	cis-3-Hexenyl benzoate	0,05
64	Pelargonic acid	1,14
65	Eugenol	0,06

Continued on next page.

Table V. (Cont.)

Peak No.	Compound	GC - Area %
66	Docosane	0,10
68	2-Heptadecanone	6,65
70	Capric acid	0,05
71	Tricosane	3,67
72	Tricosene	0,05
73	2-Octadecanone	0,06
74	Octadecanal	0,30
76	Undecanoic acid	0,10
77	Tetracosane	0,31
78	2-Nonadecanone	0,60
80	Pentacosane	3,50
81	Pentacosene	0,06
84	Eicosanal	0,28
88	Benzyl benzoate	0,08
89	Heptacosane	0,56
90	Myristic acid	0,01

Table VI: **Volatile constituents of Meadowsweet flowers**
(*Filipendula ulmaria* (L.) Maxim., Syn *Spiraea ulmaria* L.)
Closed Loop Stripping of living flowers

Peak No.	Compound	GC-Area %
11	Isoamyl alcohol	0,34
12	Limonene	0,70
16	p-Cymene	1,30
19	cis-3-Hexenyl acetate	5,53
22	cis-3-Hexenol	1,23
28	Benzaldehyde	5,42
29	Camphor	1,40
30	Linalool	0,61
33	Benzonitrile	0,85
34	Methyl benzoate	34,23
35	Ethyl benzoate	0,99
37	Salicylaldehyde	0,33
48	Benzyl alcohol	1,20
57	Anisaldehyde	1,82
59	Hexyl benzoate	0,19
67	Methyl palmitate	0,15
69	Cadalene	6,18
88	Benzyl benzoate	0,45

pin described earlier (*29, 30*) could not be confirmed. In addition to known substances such as anisaldehyde, benzaldehyde, and phenylethyl alcohol (*31*), we found a number of compounds not yet described for meadowsweet flower oil: α- and ß-pinene, limonene, trans-ß-ocimene, gamma-terpinene, lavandulol, lilac aldehyde and lilac alcohol, geraniol, geranyl acetone, hexyl benzoate, hexahydrofarnesyl acetone, cis-3-hexenyl benzoate, and some saturated and unsaturated aliphatic hydrocarbons and ketones. Especially remarkable is the occurrence of a relatively large amount of ß-damascenone which occurs in meadowsweet oil at a concentration approximately 10 times higher than in rose oil (*32*). Because of its low threshold the ß-damascenone is one of the most important constituents of rose oil and might contribute significantly to the sensory properties of meadowsweet oil.

Methyl salicylate was missing in the headspace of meadowsweet flowers which also contained only 1 % of salicylic aldehyde. But the main component of the headspace is methyl benzoate which does not occur in the steam distillate. Other substances such as cis-3-hexenyl acetate, cis-3-hexenol, benzaldehyde, linalol, benzyl alcohol and phenyl ethyl alcohol could be found in the headspace as well as in the steam distillate but in very different concentrations. Important for the odor impression of the meadowsweet flower smell are possibly benzonitrile with its almond character and anisaldehyde with its sweet floral and vanilla-like note (Table VI).

Conclusion

We have again shown remarkable differences between the analytical results obtained from living medicinal plants and from corresponding distilled oils. The headspace method again proved to be a valuable tool which for the time being mainly produces analytical data highly esteemed for creative perfumery work. We, however, should not forget classical distillation which sometimes results in artifacts of value to our industry.

Literature Cited

1. Schilcher, H. *Die Kamille, Handbuch für Ärzte, Apotheker und andere Naturwissenschaftler*; Wissenschaftliche Verlagsgesellschaft m.b.H.; Stuttgart, **1987**, 18-19, 97-98.
2. Lawrence, B. M. *Perf. Flavorist*, **1987**, *12*, 36-52, et loc. cit.
3. Kaiser, R. and Lamparsky, D. *Parf. Cosm. Arômes*, **1977**, *17*, 71-79.
4. Kaiser, R. Trapping, Investigation and Reconstitution of Flower Scents; In *Perfumes, Art, Science and Technology*; Müller, P.M., Lamparsky, D., Eds.; Elsevier Applied Science, London, New York, **1991**; 213-250.
5. Georges, G.; Fellous, R. *Parf. Cosmet. Arômes*, **1982**, *43*, 37-46.
6. Bicchi, C.; Frattini, C.; Raverdino, V. *J. Chromatogr.* **1987**, *411*, 237-249.
7. Hasebe, A.; Oomura, T. *Koryo*, **1989**, *161*, 93-101.
8. Klimes, I.; Lamparsky, D.; Scholz, E. *Helv. Chim. Acta*, **1981**, *64*, 2338-2349.

9.   Klimes, I.; Lamparsky, D. *Perf. Flavorist*, **1984**, *9*, 1-13.
10.   Thomas, A. F. *Helv. Chim. Acta*, **1981**, *64*, 2397-2400.
11.   Thomas, A. F.; Egger, J. C. *Helv. Chim. Acta*, **1981**, *64*, 2393-2396.
12.   Thomas, A. F.; Schouwey, M.; Egger, J.-C. *Helv. Chim. Acta*, **1981**, *64*, 1488-1495.
13.   Chialva, F.; Gabri, G.; Liddle, P. A. P.; Ulian, F. *J. HRC & CC*, **1982**, *5*, 182-188.
14.   Thomas, A. F.; Willhalm, B. *Tetrahedron Lett.* **1964**, *49*, 3775-3778.
15.   Zalkow, L. H.; Brannon, D. R.; Uecke, J. W. *J. Org. Chem.* **1964**, *29*, 2786 - 2787.
16.   Waller, G. R.; Frost, G. M.; Burleson, D.; Brannon, D. R.; Zalkow, L. H. *Phytochem.* **1968**, *7*, 213-220.
17.   Aboutabl, E. A.; Hammerschmidt, F.-J.; Elazzouny, A. A. *Sci. Pharm.* **1987**, *55* (4), 267-271.
18.   Derbesy, M; Touche, J.; Zola, A. *J. Ess. Oil. Res.* **1989**, *1*, 269-275.
19.   Vernin, G. *J. Ess. Oil Res.* **1991**, *3*, 49 - 53.
20.   Brunke, E.-J.;Hammerschmidt, F.-J.; Schmaus, G. *DRAGOCO-Report*, **1992**, *39*, 151-167.
21.   Epstein, W. W.; Poulter, C. D. *Phytochem.* **1973**, *12*, 737-747.
22.   Uchio, Y.; Matsuo, A.; Nakayama, M.; Hayashi, S. *Tetrahedron Lett.* **1976**, *34*, 2963-2966.
23.   Uchio, Y.; Matsuo, A.; Eguchi, S.; Nakayama, M.; Hayashi, S. *Tetrahedron Lett.* **1977**, *13*, 1191-1194.
24.   Uchio, Y. *Tetrahedron*, **1978**, *34*(19), 2893-2899.
25.   Bohlmann, F.; Zdero, C.; Faass, U. *Chem. Ber.* **1973**, *106*, 2904-2909.
26.   Boelens, M. H. *Perf. Flavorist*, **1986**, *11*, 43-63.
27.   Steltenkamp, R. J.; Casazza, W. T. *J. Agric. Food Chem.* **1967**, *15*(6), 1063-1069.
28.   Hager, H. *Handbuch der Pharmazeutischen Praxis*; *4*. Neuausgabe, Springer Verlag Berlin, Heidelberg, New York, **1973**, Vol. *4*, 997- 998.
29.   Kozhin, S. A.; Sulina, Y. G. *Rastit. Resur.* **1971**, *7*(4), 567-571.
30.   Saifullina, N. A.; Kozhina, J. S. *Rastit. Resur.* **1975**, *11*(4), 542-544.
31.   Lindemann, A.; Eriksson, P. J.; Lounasmaa, M.; *Lebensm.-Wiss. u. Technol.* **1982**, *15*, 286-289.
32.   Demole, E.; Enggist, P.; Säuberli, U.; Stoll, M.; Kovats, sz. E. *Helv. Chim. Acta*, **1970**, *53*, 541-551.

RECEIVED August 19, 1992

Author Index

Affiliation Index

Subject Index

A

Production: Donna Lucas
Indexing: Deborah H. Steiner
Acquisition: Barbara C. Tansill
Cover design: Amy Meyer Phifer

Printed and bound by Maple Press, York, PA

Other ACS Books

Chemical Structure Software for Personal Computers
Edited by Daniel E. Meyer, Wendy A. Warr, and Richard A. Love
ACS Professional Reference Book; 107 pp;
clothbound, ISBN 0–8412–1538–3; paperback, ISBN 0–8412–1539–1

Personal Computers for Scientists: A Byte at a Time
By Glenn I. Ouchi
276 pp; clothbound, ISBN 0–8412–1000–4; paperback, ISBN 0–8412–1001–2

Biotechnology and Materials Science: Chemistry for the Future
Edited by Mary L. Good
160 pp; clothbound, ISBN 0–8412–1472–7; paperback, ISBN 0–8412–1473–5

Polymeric Materials: Chemistry for the Future
By Joseph Alper and Gordon L. Nelson
110 pp; clothbound, ISBN 0–8412–1622–3; paperback, ISBN 0–8412–1613–4

The Language of Biotechnology: A Dictionary of Terms
By John M. Walker and Michael Cox
ACS Professional Reference Book; 256 pp;
clothbound, ISBN 0–8412–1489–1; paperback, ISBN 0–8412–1490–5

Cancer: The Outlaw Cell, Second Edition
Edited by Richard E. LaFond
274 pp; clothbound, ISBN 0–8412–1419–0; paperback, ISBN 0–8412–1420–4

Practical Statistics for the Physical Sciences
By Larry L. Havlicek
ACS Professional Reference Book; 198 pp; clothbound; ISBN 0–8412–1453–0

The Basics of Technical Communicating
By B. Edward Cain
ACS Professional Reference Book; 198 pp;
clothbound, ISBN 0–8412–1451–4; paperback, ISBN 0–8412–1452–2

The ACS Style Guide: A Manual for Authors and Editors
Edited by Janet S. Dodd
264 pp; clothbound, ISBN 0–8412–0917–0; paperback, ISBN 0–8412–0943–X

Chemistry and Crime: From Sherlock Holmes to Today's Courtroom
Edited by Samuel M. Gerber
135 pp; clothbound, ISBN 0–8412–0784–4; paperback, ISBN 0–8412–0785–2

For further information and a free catalog of ACS books, contact:
American Chemical Society
Distribution Office, Department 225
1155 16th Street, NW, Washington, DC 20036
Telephone 800–227–5558